INTRODUCTION

Plastics are synthetic or manufactured materials which have unique and remarkable properties. Many different plastics may be obtained by manipulating the molecules and changing the chemical combinations.

Plastics have important roles in agriculture, appliances, clothing, construction, electronics, furniture, packaging, transportation, and numerous other areas. It is hardly possible to talk about the future for plastics except in highly optimistic terms.

There exists in the field of plastics, a tremendous shortage of skilled employees and a real need for an expanded program in plastics education at all levels.

This text, INDUSTRIAL PLASTICS, explores the various areas of the plastics industry. It covers basic chemistry; the properties of the major resins; industrial processes of molding, casting, fabricating, machining, decorating, and finishing. It provides instruction for school laboratory activities.

INDUSTRIAL PLASTICS is intended for students in industrial arts and technical education who need a background of basic fundamentals. It is intended also for those now employed in the plastics industry who want to increase their knowledge of plastics, and their skills.

Ronald J. Baird
David T. Baird

Blown film production for packaging and construction industries.   (Alpine American Corp.)

INDUSTRIAL
PLASTICS

Basic Chemistry, Major Resins
Modern Industrial Processes

by

Ronald J. Baird
Professor, Industrial Technology
Eastern Michigan University
Ypsilanti, Michigan

and

David T. Baird
Assistant Professor, Plastics Technology
State University of New York
Morrisville, New York

South Holland, Illinois
THE GOODHEART-WILLCOX COMPANY, INC.
Publishers

Library of Congress Catalog Card Number 85-24844
International Standard Book Number 0-87006-563-7

123456789-86-98765

Library of Congress Cataloging in Publication Data

Baird, Ronald J.
  Industrial plastics.

  Includes index.
  1. Plastics.   I. Baird, David T. (David Taylor)
joint author.   I. Title.
TP1120.B33   1986   668.4   85-24844
ISBN 0-87006-563-7

CONTENTS

Chapter 1

THE PLASTICS INDUSTRY

Plastics is a multibillion-dollar industry which produces synthetic materials and products, many of which were never dreamed of only a few years ago. Today we would be utterly lost without the synthetic materials (artificial resins produced by chemical reaction of organic substances). Many products are made of plastics produced at less cost than was possible with natural materials obtained from the earth.

Fig. 1-2. Many parts of this garden tractor are made of plastics, including the molded hood and steering wheel. (Deere & Co.)

Fig. 1-1. This theme park structure is roofed with an amazingly strong architectural plastic coated fabric. (Du Pont Co.)

Plastics have become a part of society which accepts and often takes for granted the major role of these fantastic materials, Fig. 1-1. The automobile contains hundreds of plastics parts. The home, school, office and industry rely heavily upon plastic parts for television sets, computers, furniture and garden equipment, Fig. 1-2. Also, products such as carpeting and containers utilize plastics. Even a football player relies on plastics. If all of the plastics in the uniform were removed, the player would have little left for protection.

This is truly the age of plastics. In the future plastics will make possible things that have never been done before. New de-

Fig. 1-3. Plastics are used for many exterior parts of automobiles. This picture illustrates the use of plastics for body panels and trim.   (Cincinnati Milacron Co.)

signs, new shapes, new applications stemming from plastics' astonishing and limitless versatility will permit beauty of line and purpose that is sure to become the hallmark of the new American culture, Fig. 1-3.

THE WORLD OF PLASTICS

It might appear that the plastics industry is a manufacturing group unrelated to many other industrial organizations. However, the plastics industry relies heavily upon many other industries and is an integral part of numerous related industries.

The world of plastics depends to a considerable extent on the graphic arts. Most plastic products involve some type of communication, from printing on the product, Fig. 1-4, to engraving the mold to produce a message on the product. Many products of other materials involve the use of plastics. Metal products are often plastic coated. Pieces of wood are fastened together with plastic adhesives and coated with plastic finishes. Baseball and football fields are being covered with plastic imitation grass.

HOW PLASTICS INDUSTRY DEVELOPED

The plastics industry is one of the newest of the major industries. Since plastics are synthetic, or manufactured materials, it is easy to understand why the industry followed other material industries in development.

Fig. 1-4. The graphic arts industry plays a major role in decorating plastics products. Basketball and game signs were decorated by a printing process known as hot stamping. (Gladen Div., Hayes-Albion Corp.)

Fig. 1-5. Formerly made of metal, this traffic signal is produced from tough, water-resistant plastics. (Mobay Chemical Corp.)

Humans had depended upon wood, metal, concrete, glass and other natural materials for centuries, Fig. 1-5.

It was not until 1868 that the first plastic material was commercially produced. The need for replacing ivory for billiard balls led John Wesley Hyatt, a printer, to experiment with a new process, the reaction of camphor on cellulose nitrate. The result was a material that could be formed in sheets but was not suitable for molding. Called cellulose nitrate, this plastic later became known as "Celluloid." Celluloid was quickly adopted for many purposes. It was used for windows in early automobiles and became widely used for motion picture film. Cellulose nitrate, however, was highly flammable and was later replaced by plastics which would not easily burn.

It was not until 1909 that another synthetic material appeared commercially. In that year, Dr. Leo Baekeland announced a new resin, phenol formaldehyde, which was to become a major plastic in industry. Given the name Bakelite, this new plastics material could be molded using heat and pressure, to form high heat resistant products such as coffee pot handles, pan handles and electrical outlet plugs.

This was the beginning of a rapidly developing science of synthetic materials. As the years passed, new techniques along with new scientific discoveries enabled chemists to introduce new plastics with ever increas-

Fig. 1-6. Injection molding machine which was brought to the United States from Germany in the early 1930's. (American Hoechst Plastics Div.)

DATE	MATERIAL	TYPICAL PRODUCTS
1868	Cellulose Nitrate	Guitar Picks
1909	Phenol–Formaldehyde	Telephone Handsets, Distributor Caps
1927	Cellulose Acetate	Blister Packages
1927	Polyvinyl Chloride	Shower Curtains
1929	Urea–Formaldehyde	Lighting Fixtures
1935	Ethyl Cellulose	Cosmetic Packages
1936	Acrylic	Lighting Displays
1936	Polyvinyl Acetate	Adhesives
1938	Cellulose Acetate Butyrate	Screwdriver Handles
1938	Polystyrene	Refrigerator Parts
1938	Nylon	Gears, Brush Bristles
1939	Polyvinylidene Chloride	Food Wrap
1939	Melamine–Formaldehyde	Dinnerware
1941	Alkyd	Electron Tube Bases
1942	Polyester	Boat Hulls
1942	Polyethylene	Squeezable Bottles
1943	Fluorocarbon	Metal Coatings, Bearings
1943	Silicone	Gaskets and Sealants
1945	Cellulose Propionate	Ball Point Pens
1947	Epoxy	Molds and Tools
1948	Acrylonitrile–Butadiene–Styrene	Pipe and Pipe Fittings
1949	Diallyl Phthalate	Circuit Breakers
1954	Polyurethane	Foam Cushions
1956	Acetal	Tool Handles
1957	Polypropylene	Television Cabinets
1959	Polycarbonate	Street Light Globes
1962	Phenoxy	Bottles
1962	Polyallomer	Luggage
1964	Ionomer	Skin Packages, Safety Glasses
1964	Polyphenylene Oxide	Surgical Tools
1964	Polyimide	Bearings
1965	Parylene	Protective Coatings
1965	Polysulfone	Electronic Parts
1970	Thermoplastic Polyester	Electronic Parts
1973	Polybutylene	Pipe and Tubing
1975	Nitrile Resins	Containers

Fig. 1-7. Development of plastics materials.

ing properties. In 1927 cellulose acetate was produced. Injection molding (described on page 85) added a new dimension to production of items from this material, Fig. 1-6. Rapid development of vinyl resins followed by polystyrene and polyethylene led to an overwhelming volume of new plastics being introduced. Fig. 1-7 lists development dates of commercially important plastics of the industry. Every year new plastic materials are being discovered and created and there is continued innovation in existing plastics. Plastics are giving us an opportunity to meet our environmental needs precisely.

RESIN MANUFACTURING INDUSTRIES

The resin manufacturing industries are those concerned with converting raw materials into chemical compounds which are further processed into the various plastics resins. These resins are produced in many forms; as granules, Fig. 1-8, powders, liquids or pastes, ready to be sold to companies. Then, the resins will be processed by the companies into finished products. A large majority of these industries are made up of chemical companies, who have the specialized scientists and equipment to produce plastics, and petroleum companies, who have vast raw materials and chemical extensive laboratory experience, Fig. 1-9. Some companies purchase their chemicals and then produce resins. Others manufacture resins and also produce plastics in sheet, rod, tubing and other standard shapes.

Fig. 1-8. Most plastics resins are formulated as granules by resin manufacturers. These polyvinyl chloride granules are being loaded for shipment to product manufacturing companies.   (Conoco Chemical Div.)

MOLDING AND FABRICATING

Companies that make plastic products may be divided into two groups, those that do molding and those that do fabricating. Companies that usually do molding purchase plastics resins from a resin manufacturer and convert these materials into products by molding processes, Fig. 1-10. Plastics resins with a wide range of colors and properties are available. Some companies take orders for large quantities of a single item within the range of their equipment. Other companies operate as job shops and produce many different low volume items. In some instances companies not related to the plastics industry buy molding equipment to make products for themselves. A good ex-

ample is a company that does chrome plating of plastics products, Fig. 1-11.

Fabricating companies differ from those that do molding in that they begin with standard sheet and stock plastics materials rather than with granules or powders. Fabricators use many production techniques in manufacturing products. See Fig. 1-12.

DECORATING AND FINISHING

A number of companies are in the business of supplying materials and services to decorate or produce special finishes for plastics products. Some of this work is done by molders and fabricators themselves while other companies provide this service. Since most plastic products require some type of decorating, there are many companies closely related to the plastics industry that work in this area. See Fig. 1-13.

Fig. 1-9. A petrochemical facility, this plant converts liquified petroleum into gases which are chemically processed into polyvinyl chloride resins.   (Conoco Chemical Div.)

Fig. 1-10. These automotive wheel covers are automatically injection molded from granules in a modern production facility. (Cincinnati Milacron Co.)

Fig. 1-11. Kitchen faucets made of polyvinyl chloride have been chrome plated for protection and beauty. Applications of plastics such as this have replaced many metal parts. (Nibco, Inc.)

Fig. 1-12. These skis have been fabricated from plastics sheet material and laminated for strength and toughness. (United States Steel Corp.)

ADVANTAGES AND DISADVANTAGES OF PLASTICS

As with other materials, there are certain advantages and disadvantages of using plastics. Plastics differ from other materials in that they provide a combination of properties rather than extremes of single properties. This characteristic contributes much to their widespread usage. Plastics can be flexible, light, strong, and transparent all at the same time, Fig. 1-14. These combination properties make plastics ideal for many applications where other materials

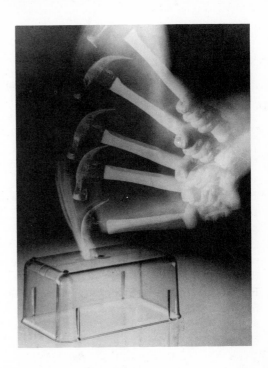

Fig. 1-14. Combining plastics properties such as toughness, strength, and clarity provide many engineering design possibilities for plastics. This polycarbonate case illustrates the value of multiple properties. (General Electric Co.)

with one outstanding property are not suitable. Fig. 1-15 illustrates some major advantages and disadvantages of plastics materials. The advantages of individual plastic types are covered in the chapter on the Major Resins of the Plastics Industry.

Fig. 1-13. The design on this melamine dinnerware was made by a process known as in-mold decorating. (Lexington United Corp.)

ADVANTAGES	DISADVANTAGES
1. Range of Color	1. Difficult to Repair
2. Insulation (Heat and Cold)	2. Objectionable Odors
3. Insulation (Electrical)	3. Unstable at High Temperatures
4. Resistance to Corrosion	4. Dimensionally Unstable
5. Weight Reduction	5. Costly
6. Ease of Processing	6. Subject to Deterioration

Fig. 1-15. Some advantages and disadvantages of plastics.

TEST YOUR KNOWLEDGE - CHAPTER 1

1. Plastics is a _____ dollar industry.
2. _____ are a large part of the plastics industry, for most plastics products involve some type of communication.
3. Of all industries, the plastics industry is one of the:

a. Smallest.

b. Oldest.

c. Newest.

4. The first commercially produced plastics material was _____ . It was introduced in the year_____. The major drawback of the first plastic was that it was highly_____ .

5. The second important plastics to be developed was_____ in the year_____. It was discovered by_____and was the first plastics suitable for_____ processes.

6. Companies that make plastic products may be divided into two groups; those that do_____and those that do_____.

7. Resin manufacturers make plastics ready for molding in these forms:

a._____.

b._____.

c._____.

d._____.

8. Plastics provide a _____ of properties rather than _____ ___ _____ properties.

9. Five major advantages of plastics are:

a._____.

b._____.

c._____.

d._____.

e._____.

10. Five major disadvantages of plastics are:

a._____.

b._____.

c._____.

d._____.

e._____.

11. In terms of properties, plastics can be _____ , _____ , _____ , and _____ all at the same time.

SUGGESTED ACTIVITIES

1. Make a list of the newest plastics products which are available in your home. Explain what properties of the material seem outstanding for each product.

2. Make a collection of all the standard shapes of plastics products you can find; sheets, rods, tubes, etc. Mount the samples on a display board and label each sample.

3. Check a late model automobile and list by name the parts you find that are made of plastics. Explain why you think plastics were used for each part.

4. Write to a plastics resin manufacturing company for the literature on the resins they produce. Make a display board using the materials you receive.

5. List the different forms in which plastics resins are produced and try to obtain samples from manufacturers.

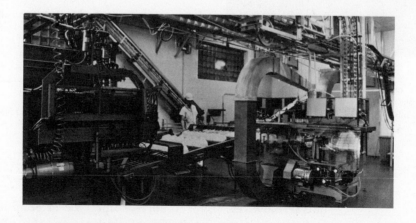

Blow molding of milk bottles in a modern dairy.   (Blowmolding Machinery Div., Hoover Universal Inc.)

Chapter 2

BASIC CHEMISTRY
OF POLYMERS

Plastics are synthetic or manufactured materials. They are formed through chemistry by taking ingredients such as coal, water, air, petroleum, limestone, salt, etc. apart and putting them together in different, desirable combinations.

Taking a substance apart enables chemists to find out (by analysis) of what it is made. By combining the parts in various ways, they may produce a completely new substance, or materially change a substance that is already available, Fig. 2-1. In the field of plastics chemistry, there have been many remarkable developments, Fig. 2-1A.

In our study of the basic chemistry of plastics, the first step is to become familiar with some of the chemical terms and concepts that will be involved. A definition of plastics should prove helpful at this point. PLASTICS is a high molecular weight synthetic product which will flow into a given form, usually when under heat and pressure.

STRUCTURE OF MATTER

All matter is made up of minute particles called ATOMS. An atom is the smallest part or component of an element which retains the properties of the element. In other words, it is the smallest division of matter, and cannot be further broken down or changed.

An ELEMENT is a fundamental substance containing atoms of only one kind. It cannot be separated into simpler substances by chemical means. For example, pure oxygen contains only oxygen atoms. However, elements can be joined to make other substances.

A COMPOUND is a distinct or pure substance formed by the chemical union of two or more elements or ingredients. A compound has properties which are different from the properties of any of the original elements. As an example, carbon unites with oxygen to form carbon dioxide. The carbon dioxide compound is different from both carbon and oxygen.

A MOLECULE is the smallest particle of an element or a compound which is capable of retaining chemical identity with the substance in mass. The elements have not changed chemically, as in compounds.

POLYMERS

Before the manufactured products called plastics are made into finished products, they are called polymers or resins. The term polymer may be defined as a chemical compound or mixture of compounds formed by polymerization (chemical reaction in which two or more molecules combine to form larger molecules). Forms in which polymers are produced include: sheet, liquid, paste, granules, powder, rope, flakes.

Fig. 2-1. Designers discuss properties of a polymer modified by chemist to meet the demands of a new product. (Cincinnati Milacron Co.)

Fig. 2-1A. Improved one-piece molded nylon bicycle wheel is lighter than a steel wheel, yet much tougher. This is a typical development through plastics chemistry. (Du Pont Co.)

CHEMICAL ELEMENTS USED IN MAKING POLYMERS

It is fortunate that the basic ingredients necessary to manufacture plastics are in abundance. As mentioned before, the raw materials commonly used are coal, air, water, wood, petroleum, limestone, and salt. Each of these is quite easily obtained. The largest volume of raw material comes from the petroleum industry, usually in the form of liquids and gases. These materials contain two or more of the six chemical elements that go into the building of a polymer ... carbon, hydrogen, oxygen, nitrogen, chlorine, and fluorine. Of these chemical elements, carbon is considered the backbone of the polymer as it usually links elements to form a polymer. The elements sulfur and silicone are sometimes used in the preparation of a few special polymers.

SYMBOLS FOR CHEMICAL ELEMENTS

Chemical elements are commonly identified by using abbreviations of chemical symbols. See Fig. 2-2.

CHEMICAL ELEMENT	SYMBOL	VALENCE BONDS	DIAGRAM SHAPE
CARBON	C	4	$-\overset{\mid}{\underset{\mid}{C}}-$
NITROGEN	N	3	$-\overset{\mid}{N}-$
OXYGEN	O	2	$- O -$
HYDROGEN	H	1	$H -$
CHLORINE	Cl	1	$Cl -$
FLUORINE	F	1	$F -$

Fig. 2-2. Common elements in plastics. Note the chemical symbols, valence bonds and diagram shapes.

Symbols are used not only to identify elements but to show the number of atoms. For example H_2 indicates that two atoms of hydrogen make up one molecule of hydrogen. Chemical formulas show both the kinds of atoms and the number of each kind in each molecule of a compound. The formula for water H_2O shows us that each molecule of water consists of two atoms of hydrogen and one atom of oxygen.

HOW ELEMENTS ARE COMBINED

To understand how plastics are formed it is necessary to know how atoms of elements are combined and the number of possibilities the chemist has for arranging them. Each atom can combine with only so many other atoms. The number depends on the combining power of the atoms involved. The combining power of an atom to join with another atom is called its VALENCE, from the Latin word meaning "power." Each element has its own valence. See Fig. 2-2.

It has been noted that in most plastics carbon is one of the elements of which it is formed. The reason for this is that carbon is one of the few elements with four valences, hence it is able to connect with at least four other atoms. Carbon easily joins with many elements and makes long chains, which is the basic structure of polymers.

Let's imagine the carbon element as a round ball with four extended arms and each of these arms as a valence bond capable of attracting another atom. One of the simplest arrangements is methane CH_4. Since a hy-

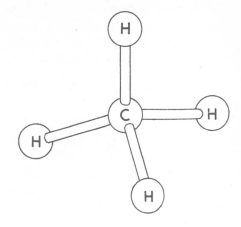

Fig. 2-3. Methane, CH_4. This shows the normal arrangement of hydrogen atoms attached to each extended valence arm of the carbon atom.

drogen atom has only one arm, the carbon atom bonds with four hydrogens and makes the only possible combination...four hydrogen atoms attached to one carbon atom, Fig. 2-3. The only way for a larger molecule to be formed would be to add more carbon atoms joined arm to arm resulting in a heavier molecule such as ethane C_2H_6, Fig. 2-4.

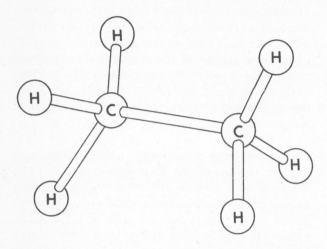

Fig. 2-4. A larger molecule, ethane, C_2H_6. This shows the connection of hydrogen atoms to the extended arms of two carbon atoms.

If this process is continued, there is actually no end to the possible length of a chain of carbon to carbon atoms. Many thousands of such atoms in a chain would still be one molecule. The adding on to the arms of carbon and other elements, is the chemist's way of building polymers. Though the elements may vary, the structure remains the same, as some arms of atoms link to form the chain, while others attach to the remaining available valence sites.

CONSTRUCTION OF A POLYMER

Items such as a colorful football helmet and a shiny plastic raincoat both began with very small building blocks (simple molecules). The molecules are the materials the chemists use. They alter their structure when necessary and group them together into long chains. By controlling the arrangement, they produce polymers of varying hardness, strength, clarity, and with other desired properties. Skill and imagination enable them to use these building blocks to produce unlimited polymers, many of which we take for granted in our daily lives.

THE MONOMER - WHERE PLASTICS BEGIN

The monomer is the starting point of all plastics. The word monomer means "one part" (derived from Greek) and refers to a material made up of small molecules all of the same size. In the monomer, the term MER refers to one repeating structural unit, Fig. 2-4A. The monomer, usually a gas or liquid, is the material the chemist uses to begin building a plastic. It is a relatively simple compound which will react, under certain conditions, to form a polymer (meaning many parts). The monomer contains all the elements which will appear in the final plastics. Fig. 2-5 illustrates some examples of monomers by showing only one molecule of each of the compounds.

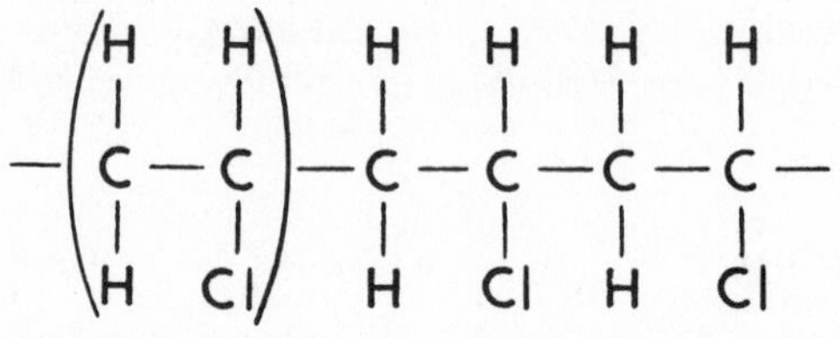

Fig. 2-4A. The mer, in parenthesis, is the repeating structural unit of any high polymer.

One of the simplest of monomers is ethylene, an odorless, colorless gas represented chemically as C_2H_4. This is a molecule containing two atoms of carbon and four atoms of hydrogen. However, it is different from ethane in that it has a double valence bond between the carbon atoms. This double valence bond is weak and can be easily broken allowing the four arms of each carbon atom to return to their normal stable

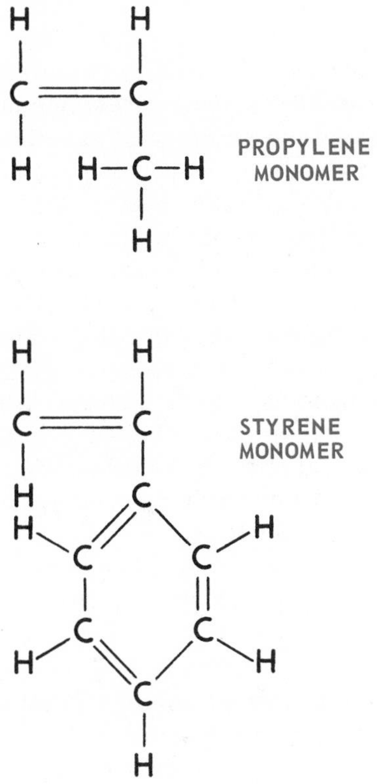

Fig. 2-5. Examples of monomers which are capable of chemical reacting to form a particular plastics type.

position, Fig. 2-6. This is the beginning of the process of changing a monomer into a polymer.

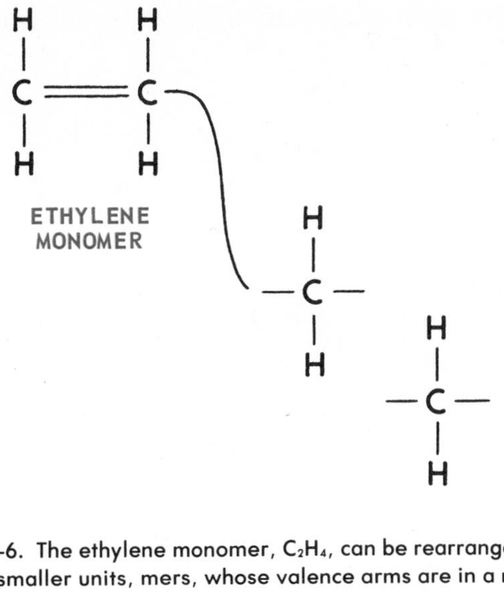

Fig. 2-6. The ethylene monomer, C_2H_4, can be rearranged to form smaller units, mers, whose valence arms are in a more stable position and ready to link up with other units.

THE PROCESS - POLYMERIZATION

In order to make a plastic material, the monomer must be put through a special type of chemical reaction. This reaction is known as polymerization. It is a complicated process in the chemical laboratory but quite simple in theory. Polymerization may be defined as a chemical reaction in which the molecules of a monomer are linked together to form large molecules whose molecular weight is a multiple of that of the original substance. During polymerization of a monomer, one of the arms of the double bond between two atoms is set free to grasp the arm of another atom which likewise has been set free. The breaking of the weak double bond to fill the arms between atoms may be caused by heat, pressure, or other chemicals. Once the atoms are freed of their double connections, they will link up in a more relaxed or stable position as a long chain, Fig. 2-7. This type of polymerization

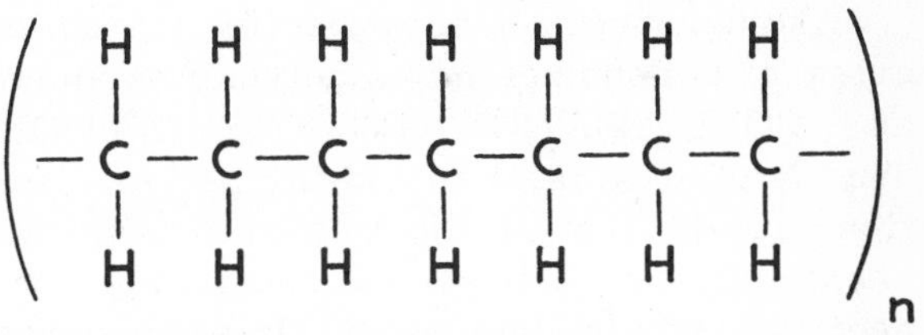

Fig. 2-7. The ethylene monomer has been changed through polymerization into a long chain molecule. The (n) represents any number of repeating units.

is known as ADDITION POLYMERIZATION. As each double bond is broken the atoms add onto one another to form a long chain, giant molecule, Fig. 2-8. In the case of the ethylene molecule, which was so small and light in weight as a GAS, the addition of thousands of atoms during polymerization caused the small molecule to become big enough and closely packed to be a SOLID plastic. It should be noted that no atomic changes take place in the monomer during polymerization. Each molecule retains all the atoms it had and no new atoms are added. The atoms simply connect to the arms of

Fig. 2-8. A long chain, giant molecule is made up of thousands of atoms connected together arm to arm.

their neighbors rather than remain alone. A plastic material formed through this process is made up of long, individual chains of molecules held together by attraction between the molecules. Any polymers formed by this process are known as THERMO-PLASTIC materials (plastics which become soft when heated and harden when cooled. The chemical composition is not changed by heating). The characteristics of thermoplastics are discussed on pages 21 and 29.

Another process of connecting the molecules of a monomer into a polymer is known as CONDENSATION POLYMERIZATION. This is quite similar to addition polymerization except that it changes the monomer chemically as it provides the needed connecting bonds between atoms. In other words, some atoms are removed from the monomer during polymerization which allows atoms from one molecular chain to grasp arms of atoms of nearby chains, Fig. 2-9. As this reaction takes place, the atoms turned loose will form separate molecules, often water,

H_2O, or other compounds, which are condenced out (separated) from the resulting plastic structure. Again, catalysts (substances which by their presence accelerate or increase the velocity of reaction between other substances), heat, or pressure cause the reaction to take place. The chemical bonding of long chain molecules, one to another, produces a three-dimensional network commonly called CROSS-LINKING. Plastics polymers formed through this reaction are generally known as THERMOSETTING materials (plastics which undergo a chemical change with heat and pressure and set into permanent shape. Thermosetting plastics cannot be softened by reheating), as described on pages 21 and 54.

THE POLYMER - THE PLASTICS MOLECULE

The final plastics material that chemists have been assembling with nature's own building blocks is the polymer - the plastics molecule. Starting with small, simple molecules from nature and the monomer, their structure is rearranged to form giant molecules known in the plastics industry as a polymer. The polymer can then be defined as a giant or high molecular weight molecule composed of many monomers linked together, Fig. 2-10. A giant molecule is actually still very small. The length of an average polyethylene molecule is about 4 billionths of an inch long, provided it is stretched out. In the

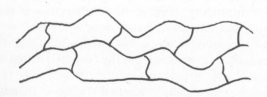

Fig. 2-9. The concept of cross-linking produces long chain molecules chemically bonded together by short chain connections as the result of condensation polymerization. Polymers of this type are strong and rigid.

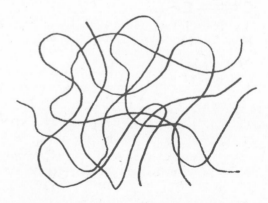

Fig. 2-10. A polymer—many long chain molecules closely associated to form a solid plastic material. These chains are not close enough to become crystalline.

plastic material, it takes the shape of a zigzag string that curls around in many directions and cannot be seen even with the most powerful microscope.

The polymers produced for commercial use, those that we see and use regularly, are manufactured as described in the preceding paragraphs. Their differences in properties, such as clarity, brittleness, waxy feel, etc., come about as a result of the arrangement, shape, and size of the molecular chains which make up the polymer and the elements from which the chain was made.

Polymers can also be produced by the polymerization of two or more monomers together. These are called COPOLYMERS. An example is styrene acrylonitrile, as the styrene and acrylonitrile monomers are reacted together to form the copolymer. Copolymerization makes it possible for the polymer chemist to taylor make plastics for the specific requirements of modern technology. Many different properties of plastics can be produced by changing the arrangement of the bifunctional units (mers) in the copolymer. By using A and B to represent the different mers, an ALTERNATING copolymer would look like this: -A-B-A-B-A-B-A-B-A-B-. They may also be joined in a random fashion, called a RANDOM copolymer, in which no recurring unit can be written: -A-B-A-A-B-B-A-B-B-B-. BLOCK copolymers are arranged in alternating groups of units: -A-A-A-B-B-B-A-A-A-B-B-B-.

There are thousands of possible arrangements of giant molecules to form different polymers, however, a knowledge of where the most important polymers were derived is necessary for this study of the plastics industry. Fig. 2-11 indicates the source of many of the commercially important plastics. The adding or subtracting of an atom of another element or a molecule of another substance produces the variety of polymers.

THERMOPLASTIC AND THERMOSETTING

Polymers may be classified into two groups, THERMOPLASTIC or THERMOSET-

METHANE
CH_4
ODORLESS, COLORLESS GAS
PHENOLIC PLASTICS   MELAMINE PLASTICS
UREA PLASTICS   ACRYLIC PLASTICS
TETRAFLUOROETHYLENE PLASTICS

ETHYLENE
C_2H_4
ODORLESS, COLORLESS GAS
RAYON PLASTICS   VINYL PLASTICS
POLYESTER PLASTICS   POLYVINYLIDENE PLASTICS
POLYETHYLENE PLASTICS

PROPYLENE
C_3H_6
ODORLESS, COLORLESS GAS
EPOXY PLASTICS   POLYPROPYLENE PLASTICS
ALKYD PLASTICS   CELLULOSIC PLASTICS

BENZENE
C_6H_6
AROMATIC, CLEAR LIQUID
STYRENE PLASTICS   POLYURETHANE PLASTICS
ABS PLASTICS   NYLON PLASTICS

Fig. 2-11. Most plastics are produced from common chemical compounds.

TING. The basis for this classification is the way in which the monomer was polymerized. You will recall that addition polymerization produces thermoplastic materials and condensation polymerization usually produces thermosetting materials. Of greater importance are the properties of these materials as final products.

Thermoplastic polymers are characterized by softening upon heating and hardening by cooling. Since the giant molecules of these materials have no strong bonds between the individual molecules, they can be softened by heat and remolded over and over again. This is an advantage in molding processes such as extrusion or injection where scrap products can be reground and molded again, Fig. 2-11A. Some of the thermoplastic materials will burn freely when exposed to an open flame while others of this group will not support combustion.

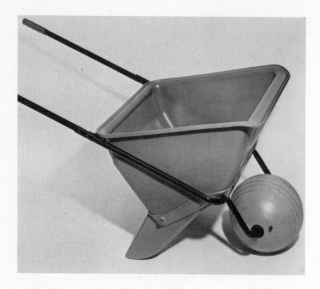

Fig. 2-11A. Thermoplastic materials, like this polyethylene wheelbarrow, are easily molded. Any scrap or defective parts can be reground and molded again. (Chemplex Co.)

chains. This refers to thermoplastic polymers as the thermosetting materials are limited in the variability of their structure.

The structure and characteristics of thermoplastics can be described by their CHEMICAL TYPE, MOLECULAR SHAPE, and MOLECULAR ARRANGEMENT.

Fig. 2-11B. Thermosetting melamine formaldehyde meets property requirements of this dinnerware by being hard, strong and highly heat resistant. (Lexington United Corp.)

The group of thermosetting polymers, numbering less than the thermoplastic group, possesses quite different characteristics. Because of the irreversible reaction by which they polymerize, they form a rigid, hard and often brittle, infusible mass. The crosslinking molecular structure with strong chemical bonds between the polymer chains causes these materials to be rigid and hard as no slippage can occur between the polymer chains. Since all the bonds are strong, when the material is heated no chain flow or softening can occur. Intensive heating of a thermoset will cause breakage of the chemical bonds resulting in a charring of the material. They are not flammable. In general, thermosetting plastics can be described as being hard, strong, and rigid, with good heat resistance, Fig. 2-11B.

BASIS FOR POLYMER STRUCTURE

It is necessary to understand why so many plastics polymers that look so much alike display such different properties. Why are some flexible and milky colored while others may be clear and hard? The basis for this understanding lies in the many arrangements the chemist can make in the molecular

CHEMICAL TYPE

The basic backbone of the plastics polymer is the carbon to carbon chain, Fig. 2-12. These chains form somewhat of a zigzag

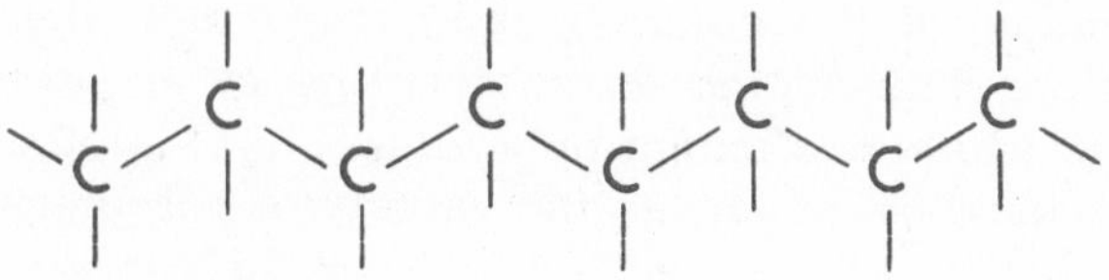

Fig. 2-12. The carbon to carbon chain is the basic backbone or ingredient of the polymer. Sites are available for side groups to attach along the chain. Most of these will be hydrogen.

three-dimensional shape. On each carbon atom there are two additional arms extended as bonding sites for other atoms. Most of these bonding sites will be occupied by hydrogen atoms but some sites can be occupied by "side groups." It is these side groups that will determine the CHEMICAL TYPE of the polymer. Some examples of side groups are shown in Fig. 2-13. A giant molecule of polystyrene, for example, would contain a benzene side group attached at every other carbon atom. The same basic structure is altered when the benzene side group is replaced by the methyl side group to form polypropylene, Fig. 2-14.

The chemical type, determined by the side group, will affect the characteristics

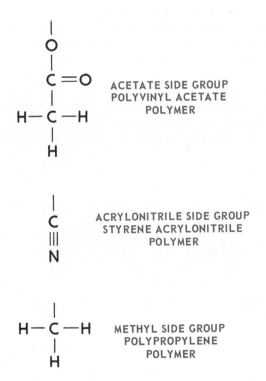

Fig. 2-13. Some examples of side groups which will determine the chemical type of the resulting polymer.

of a plastic in two ways. First, some side groups have high attractive forces for each other or other types of atoms and will cause "rigidity" in a plastic. This is due to intermolecular attraction which restricts move-

ment of the molecules. The rigidity of polyvinyl chloride, caused by the strong attraction of the chlorine atom to other atoms, is a good example of this characteristic.

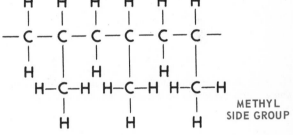

Fig. 2-14. The attachment of side groups along the carbon to carbon chain. When the benzene side group, producing polystyrene, is replaced by a methyl side group, the resulting polymer is polypropylene.

The second effect of a side group is its size. The larger the side group the more interference between molecules and an intertwining effect causing the plastic to be rigid and brittle. An example of the effect of side group size is "brittle" polystyrene with its very large benzene ring.

MOLECULAR SHAPE

The MOLECULAR SHAPE deals with the "shape" of the plastic molecule and "placement" of the side groups along the carbon to carbon backbone.

A plastic molecule can have two shapes, either "linear" or "branched," Fig. 2-15. Linear molecules, those with few or no branches, can pack very close together and cause CRYSTALLINE areas in the polymer,

Fig. 2-16. On the other hand, branched molecules with short carbon chains protruding from the main chain prevent the molecules from packing close together. This causes the polymer to be AMORPHOUS (random arrangement of molecules). Crystalline or linear plastics will be harder and less flexible than branched or amorphous plastics.

Fig. 2-17. Plastics molecules, in which the side groups are placed above and below the chain backbone, are said to be atactic.

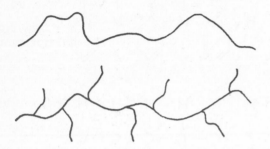

Fig. 2-15. A linear molecule, top, and a branched molecule, below, show the two basic shapes of the structure of thermoplastic polymers.

The PLACEMENT of the side groups along the carbon backbone refers to their location. If the side groups are placed at intervals around the chain the resulting molecule is called ATACTIC, Fig. 2-17. However, if the side groups are placed at the same relative site on each carbon the resulting molecule is ISOTACTIC, Fig. 2-18. From the illustration it can be seen that isotactic molecules can pack much closer

together creating crystals. Side group placement deals only with linear molecules, since branching would cancel all the effects gained by the arrangement of side groups.

MOLECULAR ARRANGEMENT

The molecular arrangement of the molecules in a polymer deals with the "size" and "weight" of the plastic molecule. For any chemical type, as the molecular size and weight increase, the values of many physical properties increase. That is, the greater the molecular weight of a polymer, such properties as strength, stiffness, and hardness increase, Fig. 2-19. Also, as molecular size and weight increase the ability of the plastic to flow decreases.

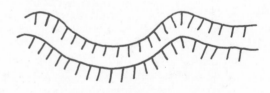

Fig. 2-18. Plastics molecules are called isotactic if side groups along chain backbone are lined up in an orderly arrangement.

RELATION OF STRUCTURE TO PROPERTIES

To show the relation of structure to properties, we will begin a simple analysis of the structure of plastics materials which causes certain physical properties to occur.

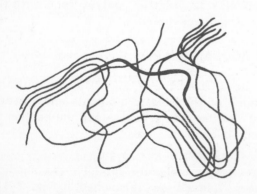

Fig. 2-16. The formation of crystals in a polymer. These are highly aligned areas due to strong forces between closely packed linear molecules.

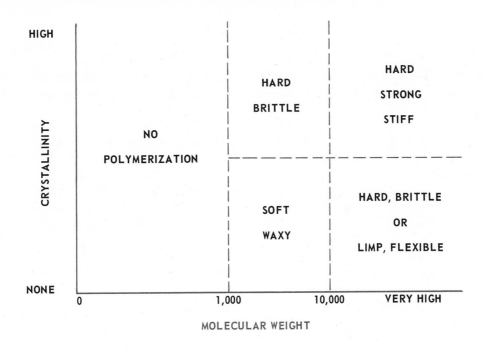

Fig. 2-19. As the molecular weight of a polymer increases, so do the physical properties. Crystallinity produces a similar effect. This chart shows how crystallinity and molecular weight affect the properties of a polymer.

Some of the principal structural features of polymers and their results in physical or chemical properties are:

CRYSTALLINITY. The tightly packed linear molecules with small side groups and isotactic arrangement which causes crystalline regions to form in the polymer:

1. High crystallinity causes cloudiness in a polymer due to light refraction, Fig. 2-19A. This shows up in linear polyethylene: about 85 percent crystalline.
2. Highly crystalline polymers have a melting point that is nearly true rather than a broad softening range.
3. Shrinkage occurs upon crystallization or cooling during molding.
4. Highly crystalline polymers can be used at higher service temperatures than most amorphous materials.
5. The higher the crystallinity the higher the density of the polymer.
6. Examples of crystalline polymers are nylon, polypropylene, tetrafluoroethylene, acetal and polyethylenes.

Fig. 2-19A. The polymer on the left shows extremely high crystallinity, average crystallinity is in center and very low crystallinity (amorphous) on right. The higher the crystallinity, the more cloudy the polymer becomes. (Kamweld Products Co., Inc.)

AMORPHOUS STRUCTURE. Structural factors which keep molecular chains far enough apart to prevent crystalline regions from forming:

1. Amorphous polymers have good clarity.

2. Less shrinkage occurs upon cooling.
3. Ease of processing is increased.
4. Large side groups tend to make polymers amorphous by keeping molecular chains apart. They may also become entwined with each other and make a polymer brittle.
5. Examples of amorphous polymers are polystyrene, acrylic, polyvinyl chloride, and cellulosics.

CHEMICAL STRUCTURE OF TYPICAL POLYMER

Several factors have been illustrated concerning the building of a polymer from the small molecule to the giant structure. Let us now look at a typical polymer, polypropylene, and see how all of these structural features fit together, and the resulting properties of the plastics material formed.

Polypropylene is made from high purity propylene gas obtained from petroleum. The propylene monomer goes through addition polymerization to form the polypropylene polymer. Fig. 2-20 illustrates the polymerization reaction.

The process is similar to that used in the production of high density polyethylene and yields a highly crystalline, linear polymer with no side branching. The lack of branches allows the molecules to pack tightly together to form the crystalline polymer. The chemist controls the process to produce an isotactic structure. Since the methyl side group is fairly large, a random arrangement along the chain would hold the molecules apart, but an ordered arrangement overcomes this effect. All of these structural features bring about the following properties of polypropylene: good heat resistance, high tensile strength, rigidity, good electrical and chemical properties, and unusual lightness with a density of 0.92.

Polymers are manufactured in many available forms, Fig. 2-21, each with unique properties developed by the polymerization chemist. The next chapter deals with the properties of individual resins based on their chemical structure.

TEST YOUR KNOWLEDGE

CHAPTER 2

1. Plastics are not materials obtained from nature so they are called _____.
2. The chemist refers to a plastic material as a polymer or a _____.
3. Define plastics in your own words by using a number of descriptive phrases.

Fig. 2-20. The polymerization reaction producing polypropylene from the propylene monomer.

Fig. 2-21. Plastics polymers are produced in many forms for further processing. Shown above are sheet, liquids, paste, granules, powder, rope, and flakes.

4. Thermoplastic materials are produced through a chemical reaction known as _____.

5. Thermosetting materials are generally produced through a chemical reaction called _____.

6. Name the six chemical elements that make up most plastics materials.

7. The chemical element considered to be the backbone of plastics is _____.

8. What are the principal raw materials from which the chemicals for making plastics are obtained?

9. The chemical reaction used to produce a plastics material involves three factors. These three factors are _____ _____ _____.

10. Is ethylene a solid, liquid, or gas?

11. Write down as many words or phrases as you can which will describe a monomer.

12. How many valence bonds or arms does carbon contain; how many for oxygen?

13. Upon heating, thermoplastic materials will _____ and then _____ when they are cooled.

14. Describe what happens when thermosetting plastics are heated to a high degree.

15. The building blocks the chemist uses to make plastics materials are _____.

16. Thermoplastic materials contain long chain molecules which are either _____ or _____ as to shape.

17. The three ways the structure and characteristics of thermoplastics can be described is by their _____ , _____ , _____.

18. How do the number of thermoplastic polymers compare with the number of thermosetting polymers available for today's industries?

19. When exposed to an open flame, _____ of the thermoplastics will burn freely.

20. The two factors that cause plastics to be

flexible, brittle, soft, or strong are _____ and _____ .

21. The two ways the side group will affect the characteristics of a plastic are _____ and _____ .

22. Cloudiness in a polymer is caused by _____ .

23. The structure of a polymer that has good clarity or transparency is said to be _____ .

24. When the side groups along the molecular chains in a polymer are attached in a regular pattern the polymer is called _____ .

25. Define the term MER as used in monomer and polymer.

SUGGESTED ACTIVITIES

1. Make a model of the monomer and the polymer of polypropylene. Make use of dowel rods, wood or plastic balls, etc. for the atoms and valence bonds.

2. Make a display board showing the principal raw materials used in the production of plastics. Attach samples of the materials.

3. Gather samples of plastic materials that illustrate different physical properties. Mount these on a chart and indicate the chemical structure that causes that particular property.

4. Secure literature from plastics manufacturing companies describing the structure of the polymers they produce.

5. Cut small samples from plastics products and drop them in a beaker of water to see if they float. Compare each sample with the density of water. Look up the density of a number of plastic polymers and compare them with the density of water.

6. Ask your instructor for a sample of polypropylene. Examine it carefully and make a list of all the physical properties you can determine through the use of your senses.

Industry photo. Extruding bubble is the first step in making polyethylene film.
(Alpine American Corp.)

Chapter 3
MAJOR RESINS OF PLASTICS INDUSTRY

THERMOPLASTIC MATERIALS

It is important to become familiar with the outstanding physical and chemical properties of thermoplastic plastics so you can make wise selections for their use. You should understand how and why these materials can be processed in certain ways. The relationship between the properties of each polymer and how these properties affect the way in which they can be molded, and why they are selected for certain products, is a key to understanding the plastics industry. Always keep in mind the following three factors as you study plastics materials:

(a) The outstanding properties of the resin.

(b) How these properties determine the ways in which it can be molded.

(c) Properties which make the resin suitable for specific products. See Fig. 3-1.

Fig. 3-1. Toughness, rigidity, lightweight and durability are properties that make plastics ideal for use in this molded office chair.
(Society of the Plastics Industry)

POLYETHYLENE RESINS

Polyethylene is the major member of a group of chemical compounds known as polyolefins. It is one of the most widely used polymers of any of the thermoplastic materials. To make polyethylene, it is necessary to use high-purity ethylene gas. The ethylene gas can be made from natural gas or a by-product of a petroleum. Through addition polymerization, the resulting polymer has the following basic structure.

$$
\begin{bmatrix}
& H & H & H & H & H & H \\
& | & | & | & | & | & | \\
- & C - C - C - C - C - C - \\
& | & | & | & | & | & | \\
& H & H & H & H & H & H
\end{bmatrix}
$$

Polyethylene is produced in two forms in terms of density. Low and intermediate density polyethylenes are produced with a relatively short chain molecular structure and a high degree of side branching. This structure provides a polymer with approximately 65 percent crystallinity. On the other hand, high density polyethylene is polymerized to form much longer linear chains with few side branches. This results in a greater density and a crystallinity range of 85 percent. The newer ultrahigh molecular weight polyethylene is of extremely high crystallinity. It contains few or no side branches and very long, tightly packed molecular chains. Polyethylene can best be described as a flexible, tough, chemically resistant, crystalline polymer, Fig. 3-2.

CHARACTERISTICS: Polyethylene appears in its natural form as a milky white, waxy feeling material. In general, as the density increases, the stiffness, hardness, strength, heat distortion point, and ability to transmit gasses increases. As density decreases, impact strength and stress crack resistance increases. Stress cracking is a surface change that polyethylene, and some other plastics, undergo when exposed to oils, gasoline and other hydrocarbons. It appears on the material as a flaky, cracked surface.

The following factors sum up the general properties of polyethylene:
1. Very tough at low temperatures.
2. Excellent chemical resistance.
3. High permeability to air and gasses.
4. Low in water vapor transmission.
5. Fairly high mold shrinkage.
6. Flexibility is good to excellent, even to -100 deg. F.
7. Weatherability is fair, can be improved by adding carbon black.
8. Excellent electrical insulating properties.
9. Easily colored in transparent (film), translucent or opaque material.
10. Odorless and tasteless, Fig. 3-3.

Fig. 3-2. The properties of high density polyethylene meet design requirements of these rugged milk delivery crates. (Arco Polymers, Inc.)

Fig. 3-3. This double-walled portable cooler was molded from odorless and tasteless polyethylene. (Alma Plastics Co.)

PRODUCT APPLICATIONS: Polyethylene not only finds many product applications due to its properties but also because of the many forms in which it is produced as a resin. It may be obtained in granules, powders, film, rod, tube and sheet form and molded through such processes as injection molding, fluidized bed coating, blow molding, extrusion, vacuum forming, casting and calendering (described in Chapter 9). Polyethylene is used for many purposes, including: containers, electrical insulation, housewares, chemical tubing, toys, freezer bags, flexible ice cube trays, snap-on lids and battery parts. Two major applications of polyethylene are films for packaging soft goods and other nonperishables and blow molded bottles. Squeeze bottles from low density polyethylene and detergent bottles from high density are typical products, Fig. 3-4.

POLYPROPYLENE RESINS

Polypropylene resins are made from propylene gas by addition polymerization. The process is similar to the production of the high density polyethylene. This yields a high molecular weight and high crystalline polymer with no side branches. The newer polypropylenes are isotactic in structure with a very high crystallinity compared to the former amorphous polymers. The long chain molecule of polypropylene with large side groups is represented in its linear form.

$$
\left[
\begin{array}{cccccc}
H & H & H & H & H & H \\
| & | & | & | & | & | \\
-C & -C & -C & -C & -C & -C- \\
| & | & | & | & | & | \\
H & H-C-H & H & H-C-H & H & H-C-H \\
 & | & & | & & | \\
 & H & & H & & H
\end{array}
\right]
$$

Polypropylene is the second resin in the family of polyolefins composed of long chain saturated hydrocarbons. The readily available source of propylene gas from the petroleum industry along with the improved properties over polyethylene make it one of the leading polymers on the industrial market.

CHARACTERISTICS: Polypropylene is produced as a molding material, as granules. In its natural state it is a fairly hard, cloudy

Fig. 3-4. High density polyethylene is used extensively for bottles and containers because of its ease of processing, toughness, and economy.   (Blowmolding Machinery Div., Hoover Universal Inc.)

white material. Because of the high degree of crystallinity in the polymer, it cannot be produced as a crystal clear material. However, when it is produced as film its clarity is similar to polyethylene film. It is one of the lightest plastics available with a density range of 0.890 to 0.905. This is often considered an advantage since, like polyethylene, it will float.

Fig. 3-5. Both case and lid of this battery are made of tough, chemical resistant polypropylene. (ICI Americas Inc.)

The improved properties of polypropylene over the same properties of polyethylene are: rigidity, flex life, heat distortion, tensile strength and stress crack resistance. The increase in these properties is due to higher crystallinity and the larger methyl group. The major properties of polypropylene are:

1. Good surface hardness and scratch resistance, Fig. 3-5.
2. Excellent dimensional stability.
3. Outstanding flex life as a hinge. Products with integral hinges are one of its most valued uses.
4. Excellent electrical properties even at high heat.
5. Some hydrocarbons will soften and swell the polymer.
6. Tough at temperatures from 105 deg. F. to 15 deg. F. but brittle below 0 deg. F.
7. Excellent resistance to water and gas vapor.
8. Good chemical resistance.
9. Serviceable above sterilization temperatures, 212 deg. F.
10. Easily colored in opaque and transluscent products.

The combination of properties of polypropylene make it an exciting material for the product designer. It is suitable for a wide variety of molding processes. It can be injection molded, blow molded, thermoformed, extruded into sheet, film, pipe, wire coatings and fibers, and processed by a number of other techniques. Polypropylene finds use in applications such as luggage cases, card files, cosmetic cases, and automobile accelerator pedals, which take advantage of its flex life and impact strength. It is used in hospital equipment because it is sterilizable, resistant to chemicals, and transparent in thin sections.

Polypropylene has also found use in products such as blown bottles, fibers for carpeting, housewares, electronic parts, aviation components, and the packaging industry. Fig. 3-6 illustrates the use of polypropylene in a pump housing.

Fig. 3-6. Glass filled polypropylene is used for this pump housing, the impeller, magnet housing, and exhaust chamber. Polypropylene was chosen because of its chemical resistance, ability to allow magnetic force to pass through it (not possible with some metals), and its self-sealing for assembly. (Fiberfil Div., Dart Industries Inc.)

These product properties will become more apparent to you as you process and test sheet material or produce molded parts.

VINYL RESINS

Vinyl resins cover a broad group of materials that range in properties from hard rigid products to soft flexible formulations. They are derived from the "vinyl" radical:

$$
\begin{bmatrix}
\begin{array}{cc}
H & H \\
| & | \\
C & = C \\
| & | \\
H &
\end{array}
\end{bmatrix}
$$

Various atoms or side groups may be attached to the vinyl radical to produce polymers of varying properties. These are illustrated in major polymers of the vinyl group as follows:

POLYVINYL CHLORIDE (PVC)

The largest polymer in the vinyl group is polyvinyl chloride. Its structure is represented by this chemical diagram:

$$
\begin{bmatrix}
\begin{array}{cccccc}
H & H & H & H & H & H \\
| & | & | & | & | & | \\
-C & -C & -C & -C & -C & -C- \\
| & | & | & | & | & | \\
H & Cl & H & Cl & H & Cl
\end{array}
\end{bmatrix}
$$

PVC is produced commercially from acetylene and hydrogen chloride and can be compounded to give almost any degree of flexibility to the final product by adding plasticizers, fillers, and stabilizers. The structure of PVC is only slightly crystalline due to its atactic molecular arrangement which provides good clarity to the material.

CHARACTERISTICS: The general characteristics of all vinyls are similar, such as good strength, excellent water and chemical resistance and unlimited color possibilities. PVC exhibits self-extinguishing characteristics, good weather resistance, electrical properties and abrasion resistance. Two factors that make PVC uniquely different from other plastics are its wide range of

Fig. 3-7. Office equipment makes use of flexible polyvinyl chloride simulated leather upholstery for chair coverings. (The B.F. Goodrich Co.)

properties and being self-extinguishing. The compounder can produce materials that are hard and rigid to those that are soft and flexible. This explains why polyvinyl chloride is used in so many product applications.

PRODUCT APPLICATIONS: Two examples of products taking advantage of the properties of PVC are flexible sheet material and rigid pipe and tubing. Calendered sheet is used for simulated leather applica-

Fig. 3-8. Toughness and flexibility of polyvinyl chloride make it an ideal choice for automotive steering wheels. (A. Schulman Inc.)

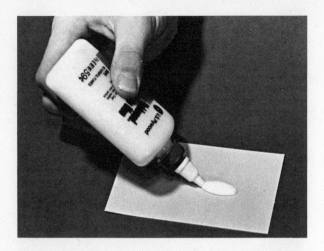

Fig. 3-9. Polyvinyl acetate in an emulsion of about 50 percent solids content has become one of the most used adhesives for wood in the forest products industry.

tions, automobile seat covers, shower curtains, and cloth or paper coated sheet for upholstery materials and raincoats. Other applications include wire coatings, chemical storage tanks, gutters and siding for houses, packaging, and blown bottles which take advantage of its good clarity and toughness. See Figs. 3-7 and 3-8.

The majority of products made of PVC are processed by extrusion, blow molding, injection, rotational, and calendering. Fluidized bed coating, foaming, transfer and compression molding are also used for some of the compounds.

POLYVINYL ACETATE

Polyvinyl acetate is prepared by the addition of acetic acid to acetylene forming the vinyl acetate monomer. Polymerization produces the polyvinyl acetate structure.

$$
\begin{bmatrix}
& H & & H & & H & & H & & H & & H & \\
& | & & | & & | & & | & & | & & | & \\
-& C & - & C & - & C & - & C & - & C & - & C & - \\
& | & & | & & | & & | & & | & & | & \\
& H & & O & & H & & O & & H & & O & \\
& & & | & & & & | & & & & | & \\
& & & C=0 & & & & C=0 & & & & C=0 & \\
& & & | & & & & | & & & & | & \\
& & & H-C-H & & & H-C-H & & & H-C-H & \\
& & & | & & & & | & & & & | & \\
& & & H & & & & H & & & & H &
\end{bmatrix}
$$

The polymer is a highly branched, atactic structure and, as would be expected, not crystalline.

Polyvinyl acetates have too low a molding temperature and are too soft to be used as molded plastics. The material is tasteless, odorless and colorless. It is used primarily as an adhesive, in an emulsion form, for wood, paper, cloth and leather, Fig. 3-9. It is also used as an extender for chewing gum, paints and flash bulb linings.

POLYVINYLIDENE CHLORIDE

A most interesting polymer for industrial and packaging uses is polyvinylidene chloride. The similarity to PVC is shown in its structure:

$$
\begin{bmatrix}
& H & & Cl & & H & & Cl & & H & & Cl & \\
& | & & | & & | & & | & & | & & | & \\
-& C & - & C & - & C & - & C & - & C & - & C & - \\
& | & & | & & | & & | & & | & & | & \\
& H & & Cl & & H & & Cl & & H & & Cl &
\end{bmatrix}
$$

It is prepared from ethylene chloride and contains an extra chlorine atom along the carbon chain. While the crystallinity is fairly low, a good transparency is obtained in film material where it finds its major use. Being odorless, tasteless and tough, with low water and vapor transmission, it is ideal as a food wrapping material such as Saran. Polyvinylidene chloride can also be formulated in both rigid and flexible forms and, being one of the most inert of all thermoplastics, can be molded, extruded and calendered. These properties make it useful in such applications as upholstery, pipe fittings and tubing, carpeting, screening material, automobile seat covers, and outdoor furniture, Fig. 3-10.

POLYVINYL CHLORIDE PLASTISOLS

Resins which are formulated as a dispersion of tiny polyvinyl chloride particles suspended in a liquid are known as PLASTISOLS. When heated, these fine particles of

Fig. 3-10. Webbing for the seat and back of this patio chair is made from polyvinylidene chloride which combines flexibility and toughness with good weatherability.
(Dow Chemical Co.)

STYRENE RESINS

Styrene resins have been in use for many years and have proven to be polymers that the industry has kept improving by modifying and copolymerizing to up-grade their properties. From the early brittle styrenes, industry now has a wide range of styrene based polymers with outstanding properties.

POLYSTYRENE

Polystyrene is a clear, odorless and tasteless polymer whose structure is long chain, linear, and amorphous.

$$\left[\begin{array}{cccccc} H & H & H & H & H & H \\ | & | & | & | & | & | \\ -C & -C & -C & -C & -C & -C- \\ | & | & | & | & | & | \\ H & \bigcirc & H & \bigcirc & H & \bigcirc \end{array} \right]$$

resin swell and later gel into a solid material. Polyvinyl chloride plastisols are used in slush, rotational, and static molding as well as in foaming and dipping processes. The major uses of plastisols are in molding flexible, soft products such as coin purses and toys. They are used also as a coating of metal parts. Plier handles, spark plug covers and drying racks make use of the coating process, Fig. 3-11.

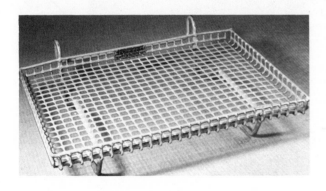

Fig. 3-11. This drive-in food service tray is made of wire coated with polyvinyl chloride plastisol. The coating is non-corrosive and provides protection to the car and food wares.

Produced from ethylene and benzene, the polymerized monomer forms a giant molecule with large side groups. It is isotactic. Scientists have worked with polystyrene structure to the extent that the polymer now produced possesses much better properties than the older resins. In general terms, polystyrene can best be described as being crystal clear, rigid, and easy to process.

CHARACTERISTICS: The wide melting range of polystyrene gives it versatility and ease of molding at various temperatures and pressures. However, there are two major disadvantages to polystyrene; its brittleness and poor chemical resistance. Brittleness is caused by the inflexibility of the molecule chains. This can be overcome by the physical addition of synthetic rubber, butadiene. This material is then known as high impact styrene and its strength is improved tremendously, making it a much more versatile plastic. The important properties of polystyrene are as follows:
1. High degree of hardness.
2. Brittle, except when modified.
3. Excellent electrical properties.

Fig. 3-12. Ease of fabrication, low mold shrinkage and reasonable cost are all properties considered when designing this insulated drink container out of polystyrene.
(Cosden Oil & Chemical Co.)

10. Poor outdoor weather resistance.
11. Normal chemical resistance is good but softens on exposure to hydrocarbons like lacquer thinner.

PRODUCT APPLICATIONS: Polystyrene products are found almost everywhere. Due to their low cost they are used for many disposable products such as picnic utensils, food containers and novelties. Typical molded products include refrigerator parts, appliance housings, furniture, Fig. 3-13, automobile interior parts, plastic optical pieces, bottles and imbedded electrical parts. Extruded polystyrene sheet is a favorite for thermoforming packaging containers, lighted indoor signs and housewares. Injection molded model airplane and car kits make extensive use of polystyrene, especially because it is easily cemented. A more recent use of polystyrene is as an expanded foam. It is outstanding for packaging delicate parts, Fig. 3-14, and as an insulation and floatation material. Picnic jugs and coolers take advantage of its outstanding insulation values.

4. Holds static electricity; picks up dust.
5. Good clarity and surface smoothness.
6. Low moisture absorption.
7. Ease of fabrication, Fig. 3-12.
8. Low cost.
9. Clear and colorless, permits outstanding colorability in transparent, transluscent and opaque shades.

Fig. 3-13. Plastic has entered the furniture industry in large volume. The solid parts of this dining set were injection molded from polystyrene.   (Shell Chemical Co.)

Fig. 3-14. These delicate electronic components are safely packaged in expanded polystyrene foam for shipping. (American Hoechst Plastics Div.)

STYRENE ACRYLONITRILE (SAN)

A copolymer of styrene, styrene acrylonitrile has improved properties of stiffness, chemical and scratch resistance and higher heat resistance. It is produced by the copolymerization of acrylonitrile and styrene to form the SAN polymer.

Due to the copolymerization, the water-white color of polystyrene is changed to a slight yellow cast in SAN. Stress crack resistance is improved. SAN finds applications in decorative panels, food packages, tumblers, Fig. 3-15, lenses, batteries, telephone parts and piano keys. However, ease of fabrication and cost lose out to gain the better properties. Styrene acrylonitrile may be molded and fabricated by processes used with polystyrene.

Fig. 3-16. Luggage injection molded of polypropylene takes advantage of its high impact strength and lightness. (Dow Chemical Co.)

ACRYLONITRILE - BUTADIENE - STYRENE (ABS)

Three monomers, acrylonitrile, butadiene, and styrene are used to produce the ABS TERPOLYMER, (TER) meaning three. It is a further way in which the properties of polystyrene can be further enhanced over SAN. Unlike the physical addition of butadiene to styrene to produce a higher impact material, ABS is a chemically polymerized plastic.

CHARACTERISTICS: ABS can be described as a rugged, tough plastic with moderately good chemical resistance and a high heat distortion point. It is one of the few thermoplastics which combines both hardness and toughness. As with most others, if it is hard, it is brittle; if it is tough, it is flexible. In its natural state ABS is a light tan colored opaque plastic with the following

Fig. 3-15. Drinking cups made of styrene acrylonitrile provide good insulation for hot or cold beverages and adequate impact strength. (Dow Chemical Co.)

$$\left[\begin{array}{c} \underset{\underset{H}{|}}{\overset{\overset{H}{|}}{C}} - \underset{\underset{\hexagon}{|}}{\overset{\overset{H}{|}}{C}} - \underset{\underset{H}{|}}{\overset{\overset{H}{|}}{C}} - \underset{\underset{C\equiv N}{|}}{\overset{\overset{H}{|}}{C}} - \underset{\underset{H}{|}}{\overset{\overset{H}{|}}{C}} - \underset{\underset{C-H}{|}}{\overset{\overset{H}{|}}{C}} - \underset{\underset{H}{|}}{\overset{\overset{H}{|}}{C}} - \underset{\underset{\hexagon}{|}}{\overset{\overset{H}{|}}{C}} - \underset{\underset{H}{|}}{\overset{\overset{H}{|}}{C}} - \underset{\underset{C\equiv N}{|}}{\overset{\overset{H}{|}}{C}} - \underset{\underset{H}{|}}{\overset{\overset{H}{|}}{C}} - \underset{\underset{C-H}{|}}{\overset{\overset{H}{|}}{C}} - \right]$$

major properties:

1. Will withstand temperatures up to 212 deg. F.
2. Low coefficient of friction.
3. Good wear and scratch resistance.
4. Resistant to most common chemicals and some hydrocarbons.
5. Good electrical properties, but flammable.
6. High hardness and rigidity.
7. Remains tough at -40 deg. F.
8. Good colorability except for transparency.

PRODUCT APPLICATIONS: The ABS polymers can be processed through most molding processes including calendering and rotational molding. They are available main-ly as powders and granules ready for processing. Typical products made of ABS are vacuum formed refrigerator door liners, luggage cases, and boat hulls. Extruded pipe and pipe fittings have found many applications. Other typical products are football helmets, a variety of automobile trim and hardware, telephone and power tool housings, tool handles, gears, and radio and television cases. See Figs. 3-16 and 3-17.

POLYCARBONATE RESINS

There are few thermoplastics that have the outstanding engineering properties of polycarbonate. In recent years it has taken its place among the most valuable resins of the industry. Polycarbonate is a member

Fig. 3-17. ABS resin was used to mold exterior panels of this trailer-camper.
(Dow Chemical Co.)

of the polyester family in that it contains carbon and oxygen atoms as the backbone of the molecular chain.

The plastics industry is using polycarbonate in direct competition with nylon and acetal as well as metals like copper, zinc, and brass.

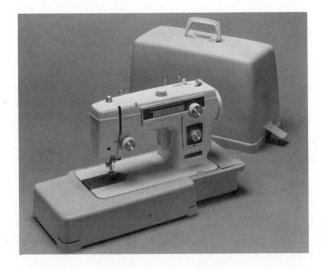

Fig. 3-18. Injection molded polycarbonate sewing machine base and cover. Polycarbonate's toughness and flexibility are causing designers to consider it for replacement of many metal parts.   (ICI Americas Inc.)

CHARACTERISTICS: One of the toughest of all plastics, polycarbonate meets many of the extreme properties of plastics with little loss in general characteristics. It is available in granular and sheet form ready for molding or fabricating. This polymer is a water-clear material, easily colored and is adaptable to a variety of molding processes including injection, extrusion, and blow molding. Five outstanding properties of polycarbonate which, in combination, separate it from other thermoplastic materials are:

1. GOOD ELECTRICAL PROPERTIES which meet the needs of electrical and electronic industries.
2. OUTSTANDING IMPACT STRENGTH which makes it useful for long service life under extreme, heavy-duty conditions, Fig. 3-18.
3. The TRANSPARENCY of polycarbonate approximates that of the light transmission of the acrylics and glass.
4. SUPERIOR DIMENSIONAL STABILITY qualifies it for use in precision-engineered components where close tolerances are required.
5. The SELF-EXTINGUISHING properties of polycarbonate make it useful in applications involving high temperature use where safety hazards may exist.

Polycarbonate also contains the average properties of many other plastics. These include good machinability, high temperature stability, good weatherability and good chemical resistance. It is attacked by some hydrocarbons and can be dissolved in ethylene dichloride.

Fig. 3-19. Olympic goggles take advantage of polycarbonate's optical clarity and high impact resistance. (General Electric Co.)

PRODUCT APPLICATIONS: The outstanding properties of polycarbonate make it suitable for applications where other thermoplastics are inadequate. Polycarbonate street lighting globes are easily blow molded and are not brittle like glass. This advantage shows up in products like protective face masks, Fig. 3-19, covers for electrical panels, electrical insulators and window panes for buildings. Other applications of polycarbonate include football and safety helmets, sunglass lenses, blow molded bottles, shoe heels, electric can openers, coffee pots, and housings for shavers, power tools and air conditioners.

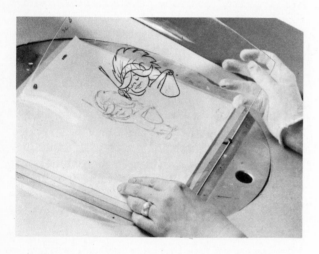

Fig. 3-20. Artist preparing a cartoon figure in ink on an overlay sheet of transparent cellulose acetate for a television commercial. Cellulose acetate provides good clarity, surface lustre, and durability.   (Celanese Plastics Co.)

CELLULOSIC RESINS

The five principal cellulosic polymers are: cellulose acetate, cellulose nitrate, cellulose propionate, cellulose acetate butyrate and ethyl cellulose. These are derived from the giant cellulose molecule obtained either from wood pulp or cotton linters. The cellulose molecule is a complex natural polymer. By substituting various chemical groups for the hydroxy groups in the molecule, the chemist is able to produce the five cellulosic plastics. Each of the polymers have a number of properties in common.

The cellulosics are some of the toughest of plastics. They have good electrical properties, are moderately heat resistant, provide a good surface lustre and can be easily colored in opaque, translucent, and transparent forms. Fast molding characteristics, solubility in some hydrocarbons, and decomposition by strong acids are other properties they share.

CELLULOSE ACETATE

Cellulose acetate is available in standard shapes (sheet, rod, tube), also as a granular molding compound ready for further processing. The natural granular material is water-white with a slight bluish tinge and is tasteless and odorless. Cellulose acetate is noted for its high impact strength and toughness. It is not suitable for outdoor weathering, but it has many useful industrial and household applications.

Cellulose acetate is usually processed by extrusion, injection molding, and transfer or compression molding. Sheet material is ideal for vacuum and pressure forming, Fig. 3-20. Extruded film is used in X-ray, photographic and tape recording equipment. Blister packaging makes use of the excellent forming qualities of cellulose acetate. Other

Fig. 3-21. Lighting fixtures molded from cellulose acetate. Cellulose acetate gives heat and electrical resistance along with toughness and good color adaptability.
(Eastman Chemical Products, Inc.)

applications include toys, safety goggles, tool handles, lamp shades, combs, and household appliances, Fig. 3-21.

CELLULOSE NITRATE

Cellulose nitrate is a rather unusual plastics material. It was the first plastic to be developed in 1868, and is still used in some applications. Due to its high degree of flammability, it cannot be satisfactorily molded by conventional processes and must be rolled into slabs and then cut into sheets. It can also be extruded cold, while it is a soft mass, into sheeting, rods and tubes. Its greatest property is toughness; one of the toughest of all plastics. Cellulose nitrate is used to some extent in industrial applications, shoe heels, and housewares. Although it is easily colored, has good water resistance, and machines well, it finds little use today because it burns so easily. Perhaps the most valuable use of cellulose nitrate is in explosives and as a pigment in lacquers.

CELLULOSE PROPIONATE

Exceptional ease of processing, stiffness, and hardness characterize cellulose pro-

pionate when compared to the other cellulosics. Its weatherability is good and it maintains a high degree of shock resistance, Fig. 3-22. Extrusion and injection molding are the main processing techniques used with cellulose propionate. These processes convert it into items such as toothbrush handles, pen and pencil barrels, steering wheels, screwdriver handles, hospital equipment, and streetlight globes, Fig. 3-23. Other applications include tubing, vacuum cleaner parts, and telephone housings.

Fig. 3-23. Breakage of streetlight globes is often decreased by replacing glass with molded cellulose propionate. (Eastman Chemical Products, Inc.)

CELLULOSE ACETATE BUTYRATE (CAB)

Cellulose acetate butyrate varies in properties from the other cellulosics in that it has better outdoor weatherability, dimensional stability, and low temperature impact strength. It has good processing characteristics and surface lustre. Conventional extrusion and injection molding techniques are used in processing along with special pow-

Fig. 3-22. Emergency highway blinker molded from cellulose propionate. (Eastman Chemical Products, Inc.)

der formulations for coating systems. Typical product applications include extruded sheet and tubing, residential mail boxes, safety glasses, tool handles, Fig. 3-24, kitchenware, home and garden equipment, Fig. 3-25, and automobile steering wheels. Vacuum formed sheet is used for outdoor signs and advertising displays. Modular units of CAB are now being produced for structural planning of new buildings and plants.

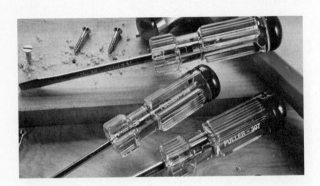

Fig. 3-24. Screwdriver handles made from cellulose acetate butyrate provide excellent impact strength, good clarity, durability and ease of processing.
(Eastman Chemical Products, Inc.)

ETHYL CELLULOSE

Ethyl cellulose is available in granules for extrusion and injection molding, as flakes for coating applications, and as cast film. Outstanding properties of the material are high impact strength at sub-zero temperatures, rigidity, good moldability, and high tensile strength. The natural color of ethyl cellulose is slightly amber yet it is easily colored from transparent to opaque. Applications taking advantage of these properties include furniture trim, doll heads, flashlight cases, and extruded tubing. Special high impact formulations make it suitable for rough usage in bowling pin bases, safety helmets, gears, and soft hammer heads, Fig. 3-26.

ACRYLIC RESINS

Acrylic (methyl methacrylate) polymers are produced from raw materials of petroleum based ethylene and propylene. The linear, amorphous polymer contains a carbon-to-carbon backbone with fairly large side groups.

$$
\left[\begin{array}{cccccc}
\text{H} & \text{CH}_3 & \text{H} & \text{CH}_3 & \text{H} & \text{CH}_3 \\
| & | & | & | & | & | \\
-\text{C} & -\text{C} & -\text{C} & -\text{C} & -\text{C} & -\text{C}- \\
| & | & | & | & | & | \\
\text{H} & \text{C}=\text{O} & \text{H} & \text{C}=\text{O} & \text{H} & \text{C}=\text{O} \\
 & | & & | & & | \\
 & \text{O} & & \text{O} & & \text{O} \\
 & | & & | & & | \\
 & \text{CH}_3 & & \text{CH}_3 & & \text{CH}_3
\end{array} \right]
$$

Many variations of the basic polymer are formed by copolymerization with other monomers, such as acrylonitrile, to produce more flexibility and other desired properties. The acrylics play a major role in the plastic industry due to their beauty, engineering properties, and ease of fabrication.

CHARACTERISTICS: The outstanding properties of acrylic resins are exceptional clarity and light transmission, Fig. 3-27, weather resistance, and colorability. The natural granular materials used for molding are water-white and hard. They can be colored in all shades from transparent to opaque and are often seen in fluorescent colors. The acrylics are also available in cast sheets which are produced from the liquid monomer polymerized between polished glass plates. Tubing, rod, and standard shapes are available through extrusion. Other properties of the acrylics are:
1. Low softening point (160 deg. F. to 200 deg. F.).
2. Good electrical properties.
3. Good cementing and heat sealing properties.
4. Excellent dimensional stability.
5. Outstanding surface lustre.
6. Low scratch resistance.
7. Strong, rigid, and good impact strength.
8. Odorless and tasteless.
9. Low moisture absorption.
10. Attacked by some chemicals such as gasoline and cleaning fluids.

PRODUCT APPLICATIONS: Although acrylic resins are not adaptable to widespread processing techniques, their use in

Fig. 3-25. This motorized mini-seeder is used for spreading small seeds such as tobacco and mustard. Cellulose acetate butyrate was selected because of its clarity, good processability, strength and weatherability. (Eastman Chemical Products, Inc.)

Fig. 3-26. Tough, molded high impact ethyl cellulose hammer heads are durable and yet do not scratch or dent precision metal parts.

product applications is extensive. Cast acrylic sheet is thermoformed into a variety of products. Exterior advertising and lighting signs, windows and canopies for boats and aircraft, skylights and translucent wall panels are a few examples, Fig. 3-28. Other applications from molding processes include lenses, radio and television parts, automotive taillights, carpeting, and containers. Acrylic rods and fibers are used to conduct light from a single source in automobile dash panels to a number of dials. Acrylics

Fig. 3-27. Moon rock displayed at the Smithsonian Institution, Washington, D.C., is protected by water clear acrylic tube. (Cadillac Plastics and Chemical Co.)

are capable of picking up light and transmitting it around curves.

NYLON RESINS

Nylon is the common name for polyamide resins. This is a group of complicated, long chain molecule polymers derived from amino and other acids.

$$\left[\begin{array}{ccccc} H & & H & O & & O \\ | & & | & \| & & \| \\ -N- & (CH_2)_6 & -N- & C- & (CH_2)_4 & -C- \end{array} \right]$$

Nylon was first introduced by the du Pont Company in 1938 and quickly became well known when it was used as fibers to weave hosiery as a replacement for silk. Because of its high impact strength, versatility, and ease of processing, nylon has become one of the leading polymers of the plastics industry. A number of modifications in production practices can be made which result in different degrees of flexibility and stiffness. This affords considerable variations, and enables specific tailor-made resins to be produced for many applications.

CHARACTERISTICS: Nylon is translucent, off-white in color, and has a high surface gloss. It can be easily colored to a broad range. Nylon is a sensitive material to process since it has a narrow melting range and must be dried, as it absorbs moisture quickly. The outstanding properties of nylon can be characterized by high heat and chemical resistance and by outstanding toughness and flexural strength, Fig. 3-29. Further properties include:
1. High abrasion resistance.
2. Low coefficient of friction.
3. Good resistance to chemicals; hydrocarbons and oils, but attacked by strong acids.
4. High moisture absorption causes dimensional change.
5. Excellent water resistance.
6. Fair electrical properties.

Fig. 3-28. This contemporary apartment house overlooks Gulf of Mexico. Its twin geodesic domes are glazed with 1/4 inch thick bronze acrylic plastic. (Rohm and Haas Co.)

Nylon is available in granules as a molding material, as a powder for coatings, and in standard sheet, rod, tubing, and fibers.

PRODUCT APPLICATIONS: Nylon continues to find many new applications in products such as electric tool housings and shaver cases where designers can take advantage of its strength in thin wall sections. Automotive interior light covers are making use of its natural translucency. Nylon can be injection molded, extruded, blow molded, and used in powder molding processes for bearings, gears, helmets, Fig. 3-30, hinges, drawer slides and rollers, combs, ship propellers, fishing lines, and textiles. More small industrial parts are made of nylon than most other plastics.

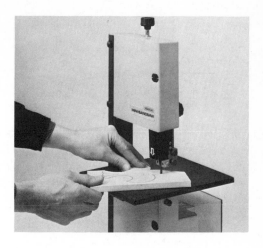

Fig. 3-29. Injection molded nylon has replaced metal parts in this bandsaw because of nylon's toughness, lightweight and good flexural strength. (ICI Americas Inc.)

Fig. 3-30. For better protection, this lightweight motorcycle helmet is made from unusually tough nylon. (Du Pont Co.)

standing electrical properties, excellent mechanical properties over a wide temperature range, high heat distortion temperature of about 375 deg. F., and good dimensional stability. These, along with low water absorption, rigidity, and strength make it suitable for a number of new applications, Fig. 3-31.

Fig. 3-31. Polyphenylene oxide was chosen for the housing of this two-cup coffee maker. Its strength and high heat distortion temperature make it ideal for this type of product. (General Electric Co.)

POLYPHENYLENE OXIDE (PPO)

Polyphenylene oxide is a polymer that is prepared through polymerization of phenol and oxygen.

$$\left[-O - \underset{CH_3}{\overset{CH_3}{\bigcirc}} - O - \underset{CH_3}{\overset{CH_3}{\bigcirc}} - \right]$$

The resulting polymer is linear in structure with a rigid phenylene oxide backbone. The natural color of PPO is an opaque medium tan with a good surface gloss. It is available as molding granules and in standard tube, rod, and sheet forms.

CHARACTERISTICS: Polyphenylene oxide is another polymer which is exciting to the design engineer. Its combination of properties are unique. Its main features are out-

PRODUCT APPLICATIONS: Although polyphenylene oxide requires molding temperatures from 550 deg. F. to 600 deg. F., it can be processed on conventional extrusion and injection molding equipment. The development of PPO led to immediate use in the electronics field due to its excellent electrical properties. Battery cases, electrical housings, Fig. 3-32, switches, printed circuits, and capacitors are a few of the uses. Other properties make it ideal for sterilizable medical and surgical instruments, food processing equipment, plumbing parts exposed to hot water, and household appliances.

ACETAL RESINS

Acetal resins are prepared from the raw material formaldehyde. Polymerization produces the acetal polymer which is a highly crystalline, unbranched, long-chain molecular structure.

$$\left[\begin{array}{c} \ \ \ \ H \ \ \ \ \ \ \ \ \ H \ \ \ \ \ \ \ \ \ H \\ \ \ \ \ \ | \ \ \ \ \ \ \ \ \ | \ \ \ \ \ \ \ \ \ | \\ -O-C-O-C-O-C- \\ \ \ \ \ \ | \ \ \ \ \ \ \ \ \ | \ \ \ \ \ \ \ \ \ | \\ \ \ \ \ H \ \ \ \ \ \ \ \ \ H \ \ \ \ \ \ \ \ \ H \end{array} \right]$$

Fig. 3-33. Lightweight fan blade, injection molded from acetal resin, is tough and shock resistant. (Cincinnati Milacron Co.)

The acetal polymers were originally produced as a replacement for many metal applications, however, the combination properties of the material also make it valuable for a variety of product requirements.

CHARACTERISTICS: Major properties of acetal are its high strength, one of the strongest of all thermoplastics; toughness; dimensional stability; and rigidity; Fig. 3-33. It has very high fatigue endurance limits, both at room and elevated temperatures. The ability of acetal to recover from loading is excellent and its impact resistance even at sub-zero temperatures is good. It is attacked by strong acids and oxidizing agents but is highly resistant to most other chemicals, including hydrocarbons, making it relatively impossible to use solvent cement. Electrical properties, even at temperatures up to 250 deg. F. are good and it maintains a low coefficient of friction. Acetal is odorless and tasteless and is quite resistant to staining

Fig. 3-32. Housings molded from polyphenylene oxide being fitted with electrical components during assembly of automotive dash units.

from foods and beverages. In its natural state it appears as a translucent to opaque white and is readily colored to a sparkling brilliance. Acetal resins are available in granules adaptable to many molding processes.

PRODUCT APPLICATIONS: Acetal resins are versatile in blow molding, extrusion, injection molding, and a number of thermoforming techniques. Typical applications resulting from these processes and the properties of acetal are automobile door handles, gears, bushings, aerosol containers, Fig. 3-34, carburetor parts, and shower heads. Acetal parts are easily fabricated by all conventional welding techniques which makes it possible to minimize parts as in a three component automotive fuel pump. Fabricating with self-tapping screws and other standard hardware is a major feature of acetal in product development.

Fig. 3-34. Aerosol bottle blow molded from acetal resin. Strength and chemical resistance make it ideal for products such as hair sprays and air fresheners.

POLYSULFONE RESINS

Polysulfone is another of the polymers produced for high temperature applications. It has an interesting chemical structure,

bringing oxygen and sulfur into the backbone of the molecular chain along with the carbon atoms.

$$\left[-\underset{\underset{CH_3}{|}}{\overset{\overset{CH_3}{|}}{C}} - \bigcirc - O - \bigcirc - \underset{\underset{O}{\|}}{\overset{\overset{O}{\|}}{S}} - \bigcirc - \right]$$

This structure combines three inherent properties in the polymer; high temperature rigidity, thermal stability, and resistance to oxidation. All of these, without the addition of modifiers, enhance its usability.

CHARACTERISTICS: The natural color of polysulfone is a clear shade of amber although it can be supplied in a variety of transparent and opaque colors. Due to the high heat resistance of the polymer it is competitive with many of the thermosetting materials. It has a heat deflection temperature of well over 325 deg. F., Fig. 3-35, and maintains its useful properties to as low as -150 deg. F. The high tensile strength of polysulfone is combined with excellent electrical properties even at temperatures up

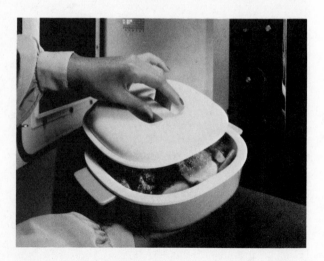

Fig. 3-35. A new concept in microwave oven cookware is this three-piece casserole made from polysulfone. The outside dish remains cool and can be handled without potholders or trivets.   (Union Carbide Corp.)

to 350 deg. F., making it extremely valuable as an electrical insulating material in high heat applications. Chemical attack by hydrocarbons and ketones will soften the surface of polysulfone; however, the overall chemical properties such as resistance to oils, acids, and salt solutions is good. Polysulfone can best be described as rigid, hard and strong with outstanding heat resistance. Absorption of moisture is high.

Fig. 3-36. Polysulfone injection molded parts for camera case. Polysulfone was selected over other materials because of its rigidity, moldability, toughness and photochemical inertness. (Union Carbide Corp.)

PRODUCT APPLICATIONS: Since polysulfone is easily processed by extrusion, injection molding, blow molding and thermoforming, the outstanding properties can be used to advantage in widespread applications, Fig. 3-36. The material is supplied mainly in granular form. Typical product applications include automobile distributor caps, where high heat and electrical resistance requirements are maintained, chemical resistant sheet and tubing, electrical circuit breakers, appliance housings, and hospital equipment that can be sterilized.

POLYURETHANE RESINS

Polyurethane resins are produced through polymerization of isocyanate and hydroxyl groups and are given the name "isocyanates" as a family of polymers.

$$
\begin{bmatrix}
& O & H & H & & O & & H & H \\
& \parallel & | & | & & \parallel & & | & | \\
-C & - N & - C & - C & - N & - C & - O & - C & - C & - O - \\
& | & | & | & | & & & | & | \\
& H & H & H & H & & & H & H
\end{bmatrix}
$$

Polyurethane is somewhat like the vinyls in that it can be prepared as a rigid molding material or a flexible, rubbery material. The major portion of polyurethane goes into the production of foamed plastics but the use of the rigid polymer is on the increase.

CHARACTERISTICS: Rigid polyurethane molding granules are an amber translucent color. They are called elastomers as they will stretch to more than twice their size and return to their original shape, Fig. 3-37. These elastomers are extremely abrasion resistant, very tough, and resist tear and shock. They are chemically resistant to almost all common chemicals including oils,

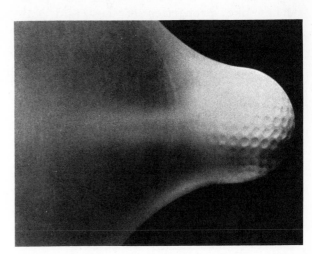

Fig. 3-37. Puncture resistance and tensile strength of polyurethane is demonstrated as this tough .003 in. thick film stops a hard-driven golf ball. (The B.F. Goodrich Co.)

Fig. 3-38. Space shuttle has a large external fuel tank with an outer insulating skin based on a polyurethane foam. (Society of the Plastics Industry)

solvents, and acids. Rigid polyurethane elastomers remain flexible down to about -40 deg. F. They have good electrical properties and high load bearing capacity, Fig. 3-38.

Foamed polyurethane has similar properties to the rigid material. It can be produced either as rigid or flexible foams in an open cellular structure. Flexible foams are good sound and energy absorbers, providing good cushioning properties.

PRODUCT APPLICATIONS: Due to the rubber-like properties of the rigid polyurethane polymers and their extreme tough-

Fig. 3-39. Tough, flexible polyurethane is molded in two sections for this automobile grille. It absorbs dents and resists scratches.

ness, they are used as solid tires on heavy equipment, printing and materials handling rolls, gaskets, bumpers and shock impact devices, Fig. 3-39, and synthetic leather. The rigid materials are injection molded, extruded, coated and cast.

Polyurethane foams are available as two component spray units for coatings, as liquids which are foamed in molds or foamed-in-place, and as slab stock for fabricating. These foamed materials are used in refrigerator insulation, sponges, crash pads, automobile and furniture cushioning, and cavity filling in boats, airplane wings, pontoons, and life jackets.

TETRAFLUOROETHYLENE RESINS (TFE)

Tetrafluoroethylene is the major member of the family of fluorocarbon polymers. It is a close relative of polyethylene, as a paraffin hydrocarbon, in which all of the hydrogen atoms are replaced with fluorine atoms. It is a long chain, linear polymer with extremely high crystallinity, 93-97 per-

$$\begin{bmatrix} F & F & F & F & F & F \\ | & | & | & | & | & | \\ -C-C-C-C-C-C- \\ | & | & | & | & | & | \\ F & F & F & F & F & F \end{bmatrix}$$

cent. The unique properties of tetrafluoroethylene and its availability in a wide range of forms make it one of the most valuable plastics on the market.

Other important members of the fluorocarbon family include fluorinated ethylene propylene (FEP), perfluoroalkoxy (PFA), ethylene-chlorotrifluoroethylene (ECTFE), ethylene-tetrafluoroethylene (ETFE) and polyvinylidene fluoride (PVDF). These resins have varying characteristics but are much alike. They have excellent chemical resistance, low coefficient of friction, and good electrical properties, along with toughness. The primary difference from tetrafluoroethylene is their ability to be processed by conventional thermoplastic equipment, such

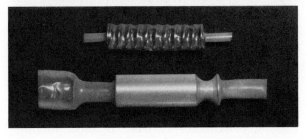

Fig. 3-39A. Top. FEP fluoropolymer heat shrinkable tubing is being applied to a steel roller by shrinking it with a heat gun while rotating. Bottom. FEP tubing completely shrunk around metal components.   (Chemplast Inc.)

as extrusion and injection molding machines. This makes them quite suitable for products such as heat shrinkable tubing, electrical components, gaskets, seals, O-rings, piping systems as well as some film applications, Fig. 3-39A.

CHARACTERISTICS: Tetrafluoroethylene has many distinguishing properties. It

Fig. 3-40. Its excellent electrical properties make tetrafluorethylene well suited for use as a wire coating. It can be used without conduit because, in a fire, it does not spread flame.   (Du Pont Co.)

is the most inert (chemically inactive) of all plastics, resisting attack from most every chemical compound known even at high temperatures. The coefficient of friction is the lowest of any known solid material. When it is exposed to temperature changes, it is still flexible at -450 deg. F. and stable up to 500 deg. F. The molecular weight of the polymer is unusually high, perhaps reaching several million. The natural color of tetrafluoroethylene is an opaque white which is readily colored. The polymer is extremely tough, possesses excellent electrical properties, Fig. 3-40, and has a waxy feeling to the touch. The low coefficient of friction also accounts for the no-stick properties displayed in many products.

Tetrafluoroethylene is a difficult material to process due to its high melting point and poor flow characteristics. Processing is accomplished by dip coating of TFE dispersions, extruding lubricated powders, and sintering similar to powdered metals. Sintering powdered TFE consists of compressing the material at room temperature and high pressure to a solid form. The formed material is then heated to about 700 deg. F. at which time fusion takes place forming a solid mass.

PRODUCT APPLICATIONS: Many industrial and consumer products make use of the extreme properties of tetrafluoroethylene in applications such as no-stick cookware, electrical insulation, chemically resistant gaskets, piston rings, bearings and tubing. Test tubes and other containers of TFE find use in chemical laboratory ware. The most commonly seen use of TFE is as a non-stick coating on rolling pins, irons, frying pans, household appliances, and many practical tools, Fig. 3-41.

IONOMER RESINS

The ionomer resins are produced from ethylene gas as the major part of the monomer. The polymer is a product of the linking together of both organic and inorganic compounds. This type of molecular structure provides a polymer with a high degree of

Fig. 3-41. Non-stick properties of tetrafluoroethylene make it ideal for tool coatings such as these. It also helps retard rust.   (Chemplast Inc.)

clarity, overcoming the effects of the crystalline formation in the plastic. Ionomer resins exhibit many of the desirable features of polyethylene, yet take on unusual properties associated with many of the amorphous polymers, Fig. 3-42.

Fig. 3-42. These golf balls made of ionomer resin combine the playability of a conventional wound-core ball and the durability of a solid ball.

CHARACTERISTICS: The outstanding properties of ionomer resins are extreme toughness, high chemical resistance, and outstanding optical properties. The natural

granular molding compound is water-white and it is odorless and tasteless. While it is easily colored, it may also be decorated by conventional metalizing, hot stamping, and painting techniques. Ionomers possess good stress crack resistance, high abrasion resistance, and their tensile strength and low temperature properties are much improved over the polyethylenes. Conventional processing of ionomers is widespread as they are easily extruded, blow molded, injection molded, and thermoformed.

TYPICAL APPLICATIONS: The clarity and formability of ionomers make them ideal for blister and skin packaging applications. They are extensively used in blow molding of bottles and containers due to their clarity and toughness. Ionomers are used also for sports equipment, toys, food packaging, automotive parts, Fig. 3-43, thermoformed containers, and tool handles. Cloth and paper are protective coated with ionomer film.

Fig. 3-43. These automobile bumper blocks are made from foamed ionomer and encased in a solid ionomer resin cover. (A. Schulman Inc.)

POLYIMIDE RESINS

Polyimide resins are unique in that, like polyesters, they can be formulated in both thermoplastic and thermosetting grades. Thermosetting polyimide is one of the most heat resistant plastics known. It can take temperatures up to 500 deg. F. (260 deg. C) for long periods and up to 900 deg. F.

(482 deg. C) intermittently. Electrical properties, wear resistance, dimensional stability and bearing qualities are all excellent. Since they are thermosetting, polyimide resins are difficult to process by conventional methods. Thus, they are normally supplied in the form of finished parts, laminates, wire enamels, adhesives and film.

Thermoplastic polyimide resins do not have quite the heat resistance of thermosets but are still very good. They followed the thermosetting types on the market and find more usage because they can be processed by most conventional means.

CHARACTERISTICS: In general, both thermosetting and thermoplastic polyimide display excellent heat resistance and electrical properties. They are specifically suited to high speed and high pressure applications, Fig. 3-44. Their insulating properties are very good as well as their high impact strength, chemical resistance and ability to be easily machined. However, they are easily attacked by strong alkaline solutions and their weathering resistance is poor.

TYPICAL APPLICATIONS: Since the polyimides are available in stock shapes, they are machinable and commonly made

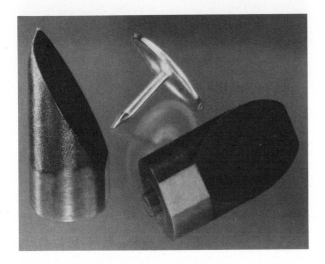

Fig. 3-44. These small "picker fingers" made from polyimide are used to strip paper from an office copier and guide it to catch tray.

into O-rings, seals, thrust washers, bearings, pump vanes and piston rings. See Fig. 3-45. The polyimides have found use in the aerospace and electronics industries as well as in office and industrial equipment. Polyimide film is used for wire insulation in electric motor windings and as an insulation for aircraft and missile wire cable.

Fig. 3-45. This is an enlarged photo of a tiny piston ring made from polyimide resin. It will withstand operating temperatures of 500 deg. F. (260 C).   (Rogers Corp.)

POLYALLOMER RESINS

Polyallomer is a member of the polyolefin family of resins as it is produced from monomers of ethylene and propylene. However, the resin differs from other polyolefin copolymers in its variation of chemical composition without change in crystallinity. This interesting material retains many of the desirable properties of both polypropylene and linear polyethylene and yet takes on unique properties of its own.

CHARACTERISTICS: Any property of a plastic which is particularly outstanding provides the designer with new dimensions for product development. Polyallomer has two such characteristics; extreme lightness in weight and unmatched "hinge" properties. In addition, it remains flexible to as low as -40 deg. F. It has favorable properties with polyethylene in ease of processing, colorability, electrical properties, abrasion resistance, and chemical resistance, Fig. 3-46. Polyallomer is produced as a granular molding resin appearing as a milky white color

readily processed by extrusion, injection molding, and thermoforming.

Fig. 3-46. Notebook, theme book, and memo pad made from polyallomer sheet resist tearing, fraying, and wash easily with a damp sponge. (Eastman Chemical Products, Inc.)

PRODUCT APPLICATIONS: Polyallomer is used for its integral hinge properties in typewriter cases, luggage, portable record player cases, and flex-top containers, Fig. 3-47. The ease of vacuum forming polyallomer makes it suitable for such items as brief case shells where the leather grain patterns maintain their shape. Extruded interior trim for automobile window panels, packaging film, and blow molded bottles are other examples of the many uses of polyallomer.

THERMOSETTING MATERIALS

Thermosetting materials differ from thermoplastics in many ways. Basically the differences are due to the chemical condensation polymerization process through which the thermosetting materials are produced that results in their special characteristics. The definition of thermosetting polymers illustrates that they are a cross-linked network of long chain molecules causing them to be rigid, strong, and infusible. This means that once they have become set or molded into a given shape they cannot be reheated and reshaped. Thermosets also have two other characteristics that are

different from thermoplastics. They are limited in the number of processes by which they can be molded and scrap resulting from molding or reject parts cannot be reused. There are, however, many advantages to thermosetting materials which will be brought out in the descriptions of the individual polymers. Most all of the thermosetting materials have fillers added to improve their properties and to extend the volume of the resin. Typical fillers are wood flour and glass fibers.

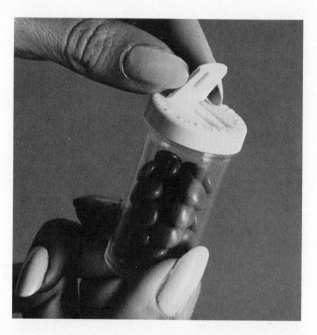

Fig. 3-47. This flex-top, snap locking safety closure makes use of polyallomer's excellent hinge properties. It was designed to discourage young children from opening medicine bottles. (Eastman Chemical Products, Inc.)

PHENOL FORMALDEHYDE RESINS

The phenol formaldehyde resins, more commonly known as phenolics, are produced from the reaction of phenol (carbolic acid) with formaldehyde in the presence of a catalyst. The resulting polymer can be obtained as a liquid for casting, bonding, coating, and impregnating, or as molding compounds as powders or flakes. Phenolic molding compounds are available only in dark colors while casting and impregnating resins can be obtained in relatively clear solutions.

duced from the polymerization of certain alcohols and acids. It is available in the form of completely polymerized polyester film, thermoplastic polyester molding granules or

CHARACTERISTICS: Phenolics can best be described as being hard, rigid, heat resistant materials that are quite brittle unless they are filled. They are seldom used in molded products without fillers which improve their toughness. Characteristics of phenolics that make them so valuable to the plastics industry are their relatively low cost, excellent insulating properties, heat resistance to 500 deg. F., and chemical inertness to most common solvents and weak acids. The dimensional stability and low moisture absorption give them an advantage in the design of precision devices.

PRODUCT APPLICATIONS: Phenolics are processed by compression and transfer molding. They are also used as liquids in laminating of veneers, fabrics, and paper, and in coating and adhesive applications. Some typical product uses of phenolics are distributor caps and coil tops, impregnated brake linings, telephones, tool housings, Fig. 3-48, and home appliance handles and parts, Fig. 3-49. Considerable volume is used as an adhesive in bonding plywood. Impregnated wood fibers are molded into salad bowls and croquet balls, while reinforced laminated sheet is used for roof panels, large ductwork, and automobile body parts. Phenolic bonding agents are used in the foundry industry for shell molding and cores.

POLYESTER RESINS

Polyester is best known as the resin combined with glass mat or chopped fibers and called fiber glass. The polymer is pro-

as a resin to which a catalyst is added to complete the curing during molding. Liquid polymer is clear and colorless. It can be filled, colored, and reinforced to meet many specifications.

Fig. 3-48. Tool housings, such as this power drill, make use of the rigidity, strength, and insulating properties of molded phenolic. (Hooker Chemicals and Plastics Corp.)

Fig. 3-49. Hard, heat resistant phenolic resin was chosen for the legs, handles and top supports of this lightweight, portable oven. (General Electric Co.)

CHARACTERISTICS: The properties of polyesters are varied as to the form in which they are processed. In general they have very good weathering and chemical resistance properties, withstanding most solvents, acids and salts. Polyesters are quite strong and tough, and depending upon the materials used for reinforcing, can be made rigid or fairly flexible. The electrical insulation properties are high along with good heat resistance to 500 deg. F. and colorability. Polyesters are available as liquid resins along with catalysts which are mixed at the time of use. Premixed molding compounds with fillers are available for immediate use. Thermoplastic polyester granules are generally used for injection molding.

PRODUCT APPLICATIONS: Polyester laminating with reinforced fibers is one of the largest commercial processes in the plastics industry. Matched die, bag, or hand lay-up molding is used to produce such familiar items as boat hulls, automobile and aircraft body components, Fig. 3-50, wash tubs and good quality luggage. Premixed resins are compression molded to produce such products as hammer handles and automotive ductwork. Injection molding of thermo-plastic polyesters has applications in pump housing, fuse cases, and automotive gears and distributor caps. Fig. 3-51 shows the use of thermoplastic polyester in typical extruded products.

Fig. 3-51. Injection molded thermoplastic polyester was used to make this shower massage. Selection of the polyester resin was based on light weight, heat strength, chemical resistance and stress cracking resistance. (General Electric Co.)

Fig. 3-50. Polyester reinforced resins of high strength, fatigue resistance, fire resistance and lightweight make an ideal outer body for this gyroplane. (Durez Div., Hooker Chemical Co.)

AMINO RESINS

Melamine formaldehyde and urea formaldehyde make up a group of plastics known as amino resins. They are produced through the reaction of either melamine or urea with formaldehyde. The molecular structure of urea formaldehyde illustrates the arrangement of the long-chain crosslinked molecule.

$$\left[\sim N - C - N - C - N - C - N - C \sim \right]$$

Melamine and urea plastics are quite similar. They are clear resins which are the easiest to color of the thermosetting materials. Both are usually filled with wood flour, glass fibers or other fillers to reduce brittleness.

CHARACTERISTICS: Amino resins are extremely hard, scratch-resistant materials, Fig. 3-52. They provide a glossy finish and retain brilliant colors in translucent and opaque shades. They are resistant to cleaning fluids, gasoline, oils and other common detergents. Aminos withstand temperatures from -60 deg. F. to continuous 180 deg. F. They will not burn or soften, even in an open flame, and have good dimensional stability and electrical properties.

PRODUCT APPLICATIONS: Melamine and urea resins are supplied as molding powders or granules for compression and transfer molding processes. Molded products of urea include switch cover plates, buttons, electric mixer housings, and radio cabinets. Typical melamine products are attractive dinnerware, Fig. 3-53, coffee makers, door knobs, and business machine housings. Both melamine and urea find extensive use as adhesives especially in the woods industries. Plywood and particle board make use of urea as a binder while melamine is used for laminated lay-ups for kitchen counters and tabletops. Aminos are also used for surface coatings on paper and fabric.

ALKYD RESINS

Alkyd resins are a group of materials derived from oil modified polyester resins.

Fig. 3-52. Molded urea formaldehyde cover plate which is easily cleaned, scratch resistant and easily decorated. (Slater Electric and Mfg. Co.)

Fig. 3-53. Stain and break resistance makes melamine resin one of the leading plastics for dinnerware. Melamine also provides attractive colors and ease of decorating. (Lexington United Corp.)

The term "alkyd" was made up from the "al" of alcohol and "kyd" representing the last syllable of acid. The condensation reaction cross-links the polyester molecules providing a hard rigid thermosetting plastic.

$$\left[\sim \underset{\underset{H}{|}}{\overset{\overset{H}{|}}{C}} - O - \underset{}{\overset{\overset{O}{\|}}{C}} \diagdown \underset{}{\overset{\overset{O}{\|}}{C}} - O - \underset{\underset{H}{|}}{\overset{\overset{H}{|}}{C}} - \underset{\underset{O}{|}}{\overset{\overset{H}{|}}{C}} - \underset{\underset{H}{|}}{\overset{\overset{H}{|}}{C}} - O - \overset{\overset{O}{\|}}{C} \sim \right]$$

The alkyds are produced as molding compounds and as liquid solutions. Approximately 90 percent of alkyd resins are used in liquid form in the protective coating field.

CHARACTERISTICS: Alkyd molding compounds have excellent dimensional stability, even at elevated temperatures, Fig. 3-54. They withstand temperatures up to 400 deg. F. for continuous use. However, their electrical properties, which are normally excellent, decline at elevated temperatures. Strong acids and bases will attack the materials but they hold up well to most organic solvents. Moisture resistance is good and they are easily colored in a wide range of opaque shades.

TYPICAL APPLICATIONS: The alkyd resins in liquid solutions are used in enamels, odorless flat wall paints, appliance and automotive finishes. As molding compounds, they are transfer and compression molded into such items as vacuum tube bases, circuit breakers, radio and television components, automotive ignition parts, and parts for electrical insulation, Fig. 3-55.

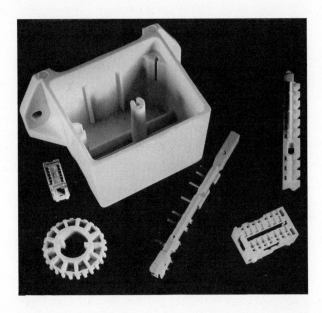

Fig. 3-55. Note assortment of television electrical parts molded from alkyd resin. (American Cyanamid Co.)

Fig. 3-54. These automotive electrical parts were transfer molded from alkyd resin. Electrical resistance and heat stability prove valuable for these applications. (Hooker Chemicals & Plastics Corp.)

EPOXY RESINS

The epoxy resins have steadily grown in their use in the plastics and related industries due to the variety of forms in which they may be processed. As thermosetting materials, they are cured or cross-linked by the addition of a hardener to the original liquid resin. The repeating molecular structure is attached to terminal molecular groups as curing takes place.

$$\left[- O - \langle \rangle - \underset{\underset{CH_3}{|}}{\overset{\overset{CH_3}{|}}{C}} - \langle \rangle - O - CH_2 - \underset{\underset{OH}{|}}{CH} - CH_2 - \right]$$

CHARACTERISTICS: Epoxy resins are formulated in such a manner that they are adaptable to many processing techniques. In each of these, the outstanding characteristics of the epoxies provide excellent chemical resistance, electrical properties, and toughness. Most epoxy products can be used continuously at temperatures up to 300 deg. F. while special formulations with fillers and additives retain their properties at continuously elevated temperatures up to 500 deg. F. They have excellent mechanical and thermal shock resistance except for some rigid formulations which are quite brittle. Weather resistance and low temperature properties are good, even to temperatures at -70 deg. F.

PRODUCT APPLICATIONS: Extensive use is made of epoxy resins in coating systems. This includes coatings for corrosion and abrasion resistance in containers, pipe and tank liners, floor and wall finishes, steel and masonry surfaces. Epoxy adhesives are very strong and are especially adaptable to metals, glass, ceramics and dissimilar materials. Molding compounds of epoxy are available with catalysts incorporated ready for compression and transfer molding into such products as pipe fittings, electrical components, and bobbins for coil winding. The epoxies are used in a manner similar to the polyesters as glass fiber reinforced lay-ups, and as laminated sheet material, Fig. 3-56. Cast epoxy is used for short run molds, tools, and jigs, being filled up to 50 percent with powdered aluminum or other inert binders. Potting and encapsulation (coating with plastic for protection) is used for electronic parts, bushings, and insulators. Other applications include printed circuit boards, boat bodies, aircraft skins, body solders and sealers.

ALLYL RESINS

DIALLYL PHTHALATE (DAP) is the most widely used polymer of the allyl plastics which are thermosetting materials developed to compete with the phenolics and amino resins. These resins are formed by the polymerization of compounds containing the following group.

$$\left[- CH_2 - CH - CH_2 - \right]$$

DIALLYL ISOPHTHALATE, a companion polymer along with DAP, established itself in the 1950's as a molding material for rocket and missile components that must withstand harsh and fast changing environments.

CHARACTERISTICS: Electrical properties of DAP include high dielectric strength, low electrical loss, and excellent insulation resistance, Fig. 3-57. The cured polymer

Fig. 3-56. Glass fibers impregnated with epoxy resin formed the outer casing for the Hercules third stage rocket motor and payload. (Union Carbide Corp.)

Fig. 3-57. The supporting forms, bases, housings, shields, and cores used for the electrical components in this undersea telephone cable amplifier are molded from diallyl phthalate. This material was specified because of its insulating properties, strength, dimensional stability, and moldability.

is odorless, tasteless, and insoluble with unusual dimension stability. Diallyl phthalate compounds are designed for continuous use at temperatures up to 350 deg. F. They have practically no moisture absorption and are highly resistant to most solvents, acids, and alkalis. Consequently they have excellent resistance to staining. DAP resins can be produced in a full range of opaque and transparent colors.

Fig. 3-58. The leading edge of this sponson (flotation unit) for an Air Force helicopter is readied for assembly to its aluminum body. It is molded from diallyl phthalate prepreg material.

PRODUCT APPLICATIONS: Diallyl phthalate resins are produced as granular molding compounds with a variety of fillers. Most molding is done by compression and transfer processes. In addition to their use in molding compounds, these resins are impregnated in glass cloth and roving (called prepregs) for reinforced plastic applications, as shown in Fig. 3-58. They are also used in insulating varnishes, sealants, and decorative laminates. Typical product applications include telephones, automobile hub caps, watch crystals, dinnerware, spacecraft parts, aircraft components, and counter tops.

SILICONE RESINS

Probably no other class of synthetic materials have found so many diverse and seemingly unrelated applications as the silicone resins. Chemically they depend upon the alternating silicone and oxygen atoms of the chain backbone with typical methyl side groups. Variations in silicones are produced by attaching other side groups along chain

$$
\begin{bmatrix}
& CH_3 & & CH_3 & & CH_3 & & CH_3 \\
& | & & | & & | & & | \\
- & Si & - O - & Si & - O - & Si & - O - & Si & - O - \\
& | & & | & & | & & | \\
& CH_3 & & CH_3 & & CH_3 & & CH_3
\end{bmatrix}
$$

and altering chain length to produce a variety of materials ranging from low viscosity fluids to semirigid solids, Fig. 3-59.

CHARACTERISTICS: Silicones are available in two basic forms; as molding compounds for transfer and compression molding, and as semiliquids for coatings and sealants. Being somewhat similar to rubber, they are stretchable up to 800 percent with excellent tensile strength. Silicones perform

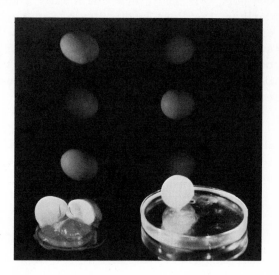

Fig. 3-59. This test illustrates the ability of silicone resins to absorb shock. An egg dropped into a silicone gel material is not broken; its fall is cushioned by silicone material.
(General Electric Co.)

Fig. 3-60. This silicone mold was made by casting RTV silicone resin around the wood pattern shown at the left. The pedestal front was reproducd by casting a filled polyester resin in the silicone mold.

as transistors and other electronic devices. Uses for molded flexible silicones include jet fuel hose, human heart valves, rubber-like gaskets and O-rings. Laminated silicones with glass cloth, asbestos, and other fibers are used in such electrical applications as transformer insulation, switch spacers, and terminal boards. The RTV (room temperature vulcanizing) silicones are used in casting flexible molds and prototype mold development, Fig. 3-60. They are available with a premixed hardener or as a separately packaged catalyst ready for easy mixing and casting. RTV silicones are also compounded with hardener incorporated for use in sealing applications, protective coatings, adhesive bonding, and as gaskets, Fig. 3-61.

satisfactorily from -140 deg. F. to 600 deg. F. under continuous use. They are highly resistant to oxidation, chemicals and oils, water, steam, and radiation. As adhesive bonds they remain tough and flexible up to 500 deg. F. They are chemically inert, odorless, tasteless, and nontoxic. Some solvents such as gasoline and benzene will cause excessive swelling of silicone compounds. The electrical properties of the silicones are good even at high temperature and moisture conditions.

PRODUCT APPLICATIONS: The molding silicone compounds are usually filled with mineral or glass fibers for use in encapsulation of electronic components such

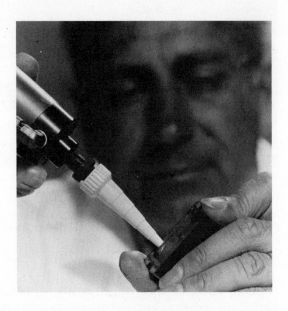

Fig. 3-61. Silicone encapsulating compounds are used in aerospace and appliance industries for sealing electronic parts.   (General Electric Co.)

TEST YOUR KNOWLEDGE - CHAPTER 3

1. Three resins that are members of the polyolefin family of plastics are _____, _____, and _____.
2. How can polyethylene best be described as to properties?
3. What is meant by stress cracking?
4. List five molded products that make use of the outstanding properties of poly-

ethylene:

a._____.

b._____.

c._____.

d._____.

e._____.

5. Does polyethylene film transmit air and gasses or does it resist them?

6. The specific gravity of polypropylene is _____. What does this factor indicate as to weight?

7. List four major ways in which polypropylene is processed into products:

a._____.

b._____.

c._____.

d._____.

8. Five individual polymers which make up the group of cellulosic plastics are:

a._____.

b._____.

c._____.

d._____.

e._____.

9. Which of the plastics has the lowest coefficient of friction of any synthetic material?

10. Most thermosetting plastics contain fillers to _____ and to _____ .

11. Two polymers based on the styrene monomer but with improved properties are _____ and _____ .

12. The most widely used thermoplastic used as an adhesive for wood and paper is _____.

13. The two largest commercial uses of polyester resins are for _____ and _____ .

14. The acetal polymers were originally developed as a replacement for many _____ applications.

15. The three special methods used in the processing of tetrafluoroethylene into products are:

a._____.

b._____.

c._____.

16. Three popular trade names for acrylic plastics are:

a._____.

b._____.

c._____.

17. Polystyrene is produced from the gas _____ and the liquid _____ .

18. The polymer that exhibits the best hinge or flex-life properties is _____ .

19. Five typical product applications which take advantage of the electrical properties of polyphenylene oxide are:

a._____.

b._____.

c._____.

d._____.

e._____.

20. The two polymers that make up the group of amino resins are _____ and _____ .

21. Acrylic plastics are noted for their:

a. Difficulty in processing.

b. Excellent resistance to high temperatures.

c. Excellent resistance to solvents.

d. High degree of light transmission.

e. None of the above.

22. The molecular weight of _____ is extremely high, perhaps reaching several million.

23. Since _____ is so highly flammable, it is seldom used for molded product applications.

24. _____ can be made up in formulations that are hard and rigid to soft, flexible types.

25. The plastic used to cast soft, flexible prototype molds is:

a. Polyester.

b. Cellulose propionate.

c. Silicone.

d. Polypropylene.

e. None of the above.

26. Polypropylene is ideal for use in hospital equipment and containers because it is:

a._____.

b._____.

c._____.

27. Properties which make polyvinylidene chloride film so suitable as a food wrapping material are:

a._____.

b._____.

c._____.

d._____.

e._____.

28. Write a brief statement to define plastisols as they relate to the polyvinyl chloride resins.
29. The general properties of natural polystyrene can best be described as:
    a. _____.
    b. _____.
    c. _____.
30. ABS is one of the few plastics which combines both the properties of _____ and _____.
31. _____ is the toughest of all plastics.
32. _____ is the common name for polyamide resins.
33. Polyphenylene oxide requires molding temperatures in the range of:
    a. 250 deg. to 300 deg. F.
    b. 350 deg. to 400 deg. F.
    c. 450 deg. to 500 deg. F.
    d. 550 deg. to 600 deg. F.

34. The natural color of polysulfone is _____.
35. The two chemical elements making up the molecular structure of tetrafluoroethylene are _____ and _____.
36. The polymer primarily produced for use in coatings and film forming material is _____.
37. Four ways in which the properties of phenolics can best be described are:
    a. _____.
    b. _____.
    c. _____.
    d. _____.
38. Three thermosetting polymers used extensively in the wood and plywood industries as adhesives are _____, _____, and _____.
39. Commercially the most used member of the group of allyl resins is _____.

SUGGESTED ACTIVITIES

1. Prepare a chart showing the different ways in which polyethylene is used for products according to the properties it displays.
2. Visit a local plastics plant (if available) and write a report on the products they manufacture, the various plastics they use, and the processes they are using to produce such products.
3. Write to a major manufacturer of plastics polymers to secure literature and specifications on the plastics they manufacture. Report to the class the properties and recommended uses of each of the polymers they sell.
4. From current literature, make a listing of the price per pound of the leading plastics. According to the specific gravity of each polymer, make a comparison of price per pound of a similar product made of each resin.
5. Make a collection of small, natural color samples of polymers in granular or powder form.
6. Prepare a display illustrating the wide variety of colors and textures in which plastics products can be produced.
7. Make a list of the outstanding properties of plastics in general that make them so valuable for commercial products.
8. Write a brief report explaining why thermosetting and thermoplastic polymers are seldom used in the same processing equipment.
9. Make a bulletin board display of plastic polymer ads.
10. Design a poster listing trade names and chemical names of plastics used in popular products such as carpeting, clothing, and housewares.
11. Secure a number of different throw-away plastic containers. Make a list of the plastic property requirements necessary to make each container function properly.

Chapter 4
LABORATORY ANALYSIS
OF RESINS

The purpose of plastics resin analysis in the laboratory is to provide a basis for identification testing. It also provides an opportunity to run tests on plastics samples which will bring about a better understanding of the chemical and physical properties of resins.

There are so many plastics on the market, it is difficult to identify all of the individual polymers. However, some polymers

Fig. 4-2. Spraying sheet metal mold with release makes the plastic separate easily.

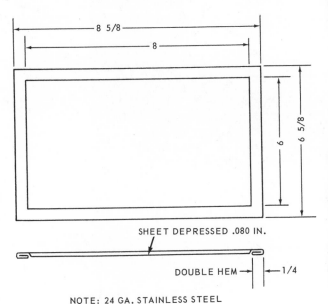

SHEET DEPRESSED .080 IN.

DOUBLE HEM — 1/4

NOTE: 24 GA. STAINLESS STEEL

Fig. 4-1. This sheet metal mold made from 24 ga. stainless steel has a double hem on each edge. After forming, a 6 x 8 in. steel block at least .080 in. in thickness is centered on the sheet and the mold is depressed in a hydraulic press.

Fig. 4-3. One-half cup of granular ABS is poured onto the mold surface.

Fig. 4-4. A dowel rod is used to spread the granules evenly over the mold surface.

Fig. 4-5. The stainless steel cover plate, sprayed with mold release, is gently placed over the mold.

are easy to identify. The same is true for many of the physical properties of plastics. Some can be tested with ease and relatively simple equipment, while others demand equipment seldom available in the school laboratory. Many of the properties of plas-

Fig. 4-6. The mold is placed in the heated press and checked to see that no granules are on the mold rim.

tics become better understood even while preparing the resin samples.

PREPARATION OF RESIN SAMPLES

Samples of some plastics are easy to find (throw-away containers and the like) but it is usually necessary to prepare your own samples of the polymers to be tested. Sheet material can be made from granules by using a simple sheet metal mold, Fig. 4-1, and a heated platen hydraulic press.

The mold should be lightly coated with a spray or other mold release, Fig. 4-2, and the press platens heated to the average molding temperature of the particular plastic sample being prepared. Molding temperatures can be located in the rear reference tables. A one-half cup measure of granules is poured into the mold, Fig. 4-3, and spread evenly over the mold surface, Fig. 4-4. The stainless steel cover plate is placed over the mold, Fig. 4-5. The sandwich is inserted into the hydraulic press, Fig. 4-6, and the pressure is brought up to just a few pounds per square inch for about one minute. During this soaking period, the granules will become molten enough that when a pressure of about 100 psi is applied they will flow to fill out the mold cavity. A

Fig. 4-8. Gently separate the cover plate and the mold from the newly formed sheet.

slight flash around the edges of the mold indicates complete flow has been obtained. Release the pressure. Using asbestos gloves, take the mold and cover plate from the press and quench in cold water, Fig. 4-7. Remove the mold from the water when it is cool, separate the sample sheet from the mold, Fig. 4-8, and dry all parts with a soft cloth, Fig. 4-9. The mold is then ready to be filled

Fig. 4-7. The hot mold is slowly cooled in a tank of water. Care should be taken in handling the hot mold.

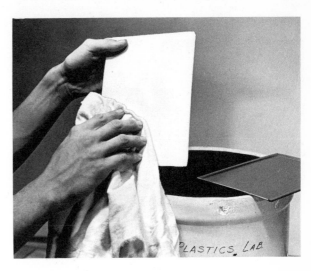

Fig. 4-9. The plastic sheet, mold, and cover plate should be carefully dried before further use.

Fig. 4-10. Molded plastic sheet being cut into 3/4 in. wide test strips on the circular saw with a special safety fixture made of hardwood.

again and returned to the press to form another sample sheet.

Plastic sheet material can be cut into 3/4 in. wide strips by sawing or by use of a paper cutter. Rigid, hard materials should be cut on a saw, Fig. 4-10. Details for construction of the safety fixture used for sawing hard plastic sheet are shown in Fig. 4-11. The softer, more flexible plastics will become gummy when sawed and can be cut easier on a paper cutter, Fig. 4-12. These strips of sample sheet plastics, Fig. 4-13, are then ready for identification testing, mounting for classification of properties, and physical-chemical tests. Full size sheets may also be prepared for vacuum forming and other processing methods. A resin flow pattern is shown in Fig. 4-14.

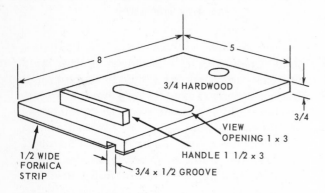

Fig. 4-11. Working drawing of circular saw safety fixture.

Fig. 4-13. Cut strips of polystyrene, polyvinyl chloride, and cellulose acetate ready for testing.

Fig. 4-12. Softer plastic sheet such as polyallomer may be cut into 3/4 in. strips using a paper cutter.

IDENTIFICATION TESTING OF RESINS

The most common procedure used to identify individual polymers is the burning test. Although a burning test is not always accurate, due to fillers and alloys which may change the characteristics of the polymer, it is satisfactory for the majority of plastics on the commercial market. Chemical solvent tests, which are more accurate, are too dangerous to be performed in a

Fig. 4-14. The flow pattern of the molded granules is obtained by thoroughly mixing equal parts of two different colored resins.

Fig. 4-16. Burning characteristics of the polymer are observed as the resin burns and drips over a metal drip sheet. Your hands should be kept well away from burning or dripping plastic.

school laboratory. These should be attempted only by experienced scientists in the plastics industry since they involve the use of hot acids and other chemicals that must be handled with extreme caution.

BURNING TEST

The first step in the identification of a plastic material is to determine if it is a thermosetting or thermoplastic resin. The best way to determine this is to heat a glass rod over a Bunsen burner flame and press it against the sample. If it softens or melts, it must be thermoplastic. Another method is to place a sliver of the sample in a test tube and heat until it darkens and decomposes (thermosetting) or melts (thermoplastic).

Most plastics have definite characteristics when exposed to flame. These are

Fig. 4-15. A test strip is brought up to the edge of the flame of the alcohol lamp until it begins to burn.

Fig. 4-17. Flame of the test sample has been blown out. Plastic is identified by odor of fumes from extinguished flame.

POLYMER BURNING IDENTIFICATION CHART

POLYMER	EASE OF LIGHTING	SELF EXTINGUISHING	ODOR	NATURE OF FLAME	BEHAVIOR OF MATERIAL
Acetal	Moderate	No	Formaldehyde	Clean blue flame, no smoke	Melts, drips, drippings may burn
Acrylic	Readily	No	Fruit-like	Blue flame; yellow top, spurts	Softens, usually no drip, little char left
Acrylonitrile Butadiene Styrene	Readily	No	Characteristic[1]	Yellow flame; black smoke	Softens, drips, chars
Cellulose Acetate	Readily	No	Acetic acid, burned sugar	Dark yellow flame; some sooty black smoke	Melts, drips, drippings continue to burn
Cellulose Acetate Butyrate	Moderate	No	Rancid butter	Dark yellow flame with blue edges, some black smoke	Melts, drips, drippings continue to burn
Cellulose Nitrate	Very readily	No, burns fast	Sharp	White flame; very rapid	Material burns completely
Cellulose Propionate	Readily	No	Fragrant	Blue flame; yellow top, sparks, some black smoke	Melts, drips, drippings continue to burn
Diallyl Phthalate	Difficult	Yes	Characteristic[1]	Yellow; black smoke	Softens, chars
Epoxy	Readily	No	Characteristic[1]	Yellow; spurts black smoke	Chars
Ethyl Cellulose	Readily	No	Burned sugar	Yellow flame; blue top and edges	Melts, drips, drippings continue to burn
Ionomer	Readily	No	Burning paraffin	Yellow-orange flame; blue edges, slight black smoke	Melts, bubbles, drips and burns, chars
Melamine Formaldehyde	Difficult	Yes	Ammonia and formaldehyde	Light yellow	Swells, cracks, turns white at edges
Nylon	Moderate	Yes	Burned wool	Blue flame; yellow top	Melts, drips, froths
Phenolic, molded	Very difficult	Yes	Burned fabric, phenolic	Yellow; a little black smoke, sparks	Cracks badly, chars, swells
Polyallomer	Readily	No	Stringent paraffin	Yellow flame; blue bottom edge, black smoke	Melts clear, spurts, drippings burn
Polycarbonate	Difficult	Yes	Sweet carbon odor	Yellow flame; dense black smoke, carbon in the air	Softens, spurts, chars, decomposes
Polyester	Moderate	No	Burning coal	Yellow; black smoke, burns steadily	Softens, no drips, continues to burn
Polyethylene	Readily	No	Burning paraffin	Blue flame; yellow top	Melts, drips, drippings may burn, swell
Polyphenylene Oxide	Moderate	No	Sweet paraffin	Yellow-orange flame; dense black smoke, carbon in the air	Softens, spurts, chars, decomposes
Polypropylene	Readily	No	Burning paraffin	Blue flame; yellow top, some white smoke	Melts, swells, drips
Polystyrene	Readily	No	Illuminating gas	Orange-yellow flame; dense smoke, clumps of carbon in air	Softens, bubbles
Polysulfone	Readily	No	Pungent sulfur odor	Yellow-orange flame; black smoke, sparkles, carbon in air	Softens, chars, decomposes
Polyurethane	Readily	No	Faint apple odor	Light yellow spurting flame; slight black smoke	Melts, drips, drippings burn
Polyvinyl Acetate	Readily	No	Acetic acid	Dark yellow flame; spurts, black smoke, carbon in the air	Softens
Polyvinyl Chloride	Difficult	Yes	Hydrochloric acid, chlorine	Yellow flame; spurts, green on edges, spurts green and yellow, white smoke	Softens
Polyvinylidene Chloride	Very difficult	Yes	Chlorine	Yellow flame; green on edges, spurts green smoke	Softens, chars, leaves ash
Styrene Acrylonitrile	Readily	No	Illuminating gas and acrylonitrile	Yellow flame; heavy black smoke, some carbon in air	Melts, bubbles, chars more than styrene
Tetrafluoroethylene	Will not ignite	Yes	Very little odor	Yellow; green near base	Melts, bubbles, slight charring
Urea Formaldehyde	Difficult	Yes	Strong pancake odor	Pale yellow flame; light greenish-blue edge	Swells, cracks, turns white at edges

[1] Odors are difficult to describe, but recognizable — use controls. (Known sample)

Fig. 4-18. Polymer Burning Identification Chart.
(Du Pont Co.)

flammability, color and nature of flame, presence or absence of smoke, melt behavior (such as dripping or swelling), and odor.

Use one of the strips of plastic prepared from the sample sheet material or cut a small strip of plastic about 3 in. long, if possible, from an unknown product. Grasp one end of this piece, using pliers or tweezers and bring the other end into the flame of an alcohol lamp or Bunsen burner. Always be careful to avoid drips of molten plastic. Hold the strip just to the edge of the flame until it takes fire, or for 10 seconds, Fig. 4-15. Withdraw the strip and, holding it over a drip sheet, Fig. 4-16, note the character of the flame, drippings, smoke, and possibility of carbon in the air. After extinguishing the flame, cautiously smell any odor produced, Fig. 4-17. Care must be taken in smelling, since some polymers may evolve toxic substances. If such danger is suspected, smell only for a very short time and indirectly by waving a hand gently through the vapors. Record your observations. These tests should be carried out in a well ventilated room.

A summary of burning behavior is found in the Polymer Burning Identification Chart, Fig. 4-18. To use this chart, first note whether the plastic burns, and if it does, whether it continues to burn after being re-moved from the flame or if it is self-extinguishing. Locate the correct grouping on the left side of the chart, and search for a more precise characterization of the burning properties in line with your data.

Although odors from heating a plastic in a test tube are similar to those encountered in the burning test, they are much stronger and more distinctive. Identification by odor may frequently be accomplished by this test when it has been indefinite in the burning test. When a sample has been tentatively identified, a further check can be made by comparison of odor with one selected from a known set of test tube samples, Fig. 4-19. When heating a plastic in a test tube, the same safety precautions should be taken as in the burning test.

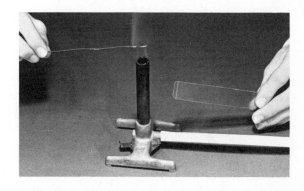

Fig. 4-20. Heating a copper wire until it turns bright red.

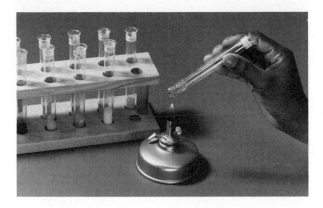

Fig. 4-19. Comparison of odor from a known test tube sample will often aid in the identification of an unknown specimen. The known sample is gently heated to emit vapors.

It may also be helpful to determine whether the polymer contains chlorine, for its presence would narrow the field of choice considerably. Heat a copper wire in a Bunsen flame until it glows, Fig. 4-20. Touch the wire to the polymer in such a way as to retain some of the polymer on the wire, and then return the wire to the flame, Fig. 4-21. A green color flame indicates the presence of chlorine. Care must be taken to do this well away from the face. Fluorocarbon polymers, vinyl chloride, vinyl chloride--vinyl acetate, and vinylidene polymers give positive results in this test.

Fig. 4-21. Copper wire coated with burning plastic emits a green flame showing the presence of chlorine in the test sample.

PHYSICAL PROPERTIES OF POLYMERS

The physical properties of polymers are the determining factors the design engineer must use in selecting a polymer with those properties which most nearly meet the requirements of a product. See Fig. 4-22. Among the major physical properties that are tested for plastics are the weight (specific gravity), hardness, flexibility, tensile strength, flowability, moisture content and impact strength. A number of these properties may be tested in the laboratory using sample materials.

SPECIFIC GRAVITY DETERMINATION

First, simply drop the plastic sample in water. If it floats, the unknown sample may be polyethylene, polypropylene, or polyallomer. If it sinks, it may be any of the plastics whose specific gravity is greater than that of water (1.00).

To obtain accurate values, this procedure must be carefully followed . . . small errors may cause large variations in the results:

The specific gravity (weight of a specimen of plastics compared to the weight of an equal volume of water) at 73 deg. F. can be determined from void free samples sawed to about 1/4 x 3/4 x 1 in. Saw marks and dirt or grease should be removed in order to avoid trapped air when immersing the sample.

Weigh the sample in air and in water. To weigh the sample in air, attach a thin wire to the sample and join the other end of the wire to the hook on one end of the pan supports of a laboratory balance. Record the weight of the sample and wire, and of the wire alone. To weigh in water, Fig. 4-23, attach the wire to the balance hook over a beaker of water and weigh the sample and

Fig. 4-22. Laboratory testing is an important part of the plastics industry. Here a technician runs a polymer weathering test on three cyclic ultra-violet weathering testers. (Molded Fiber Glass Companies)

Fig. 4-23. Arrangement for weighing a plastic sample in water with a pan balance to determine the specific gravity.

wire in the water. Then weigh the wire alone in the water. The water should be about 71 to 75 deg. F. Again record the weight of both wire and sample and of the wire alone in the beaker of water.

MEASUREMENTS

Weight of sample and wire in air . . . = B
Weight of wire in air . . . . = A
Weight of wire plus sample
   in water . . . . = D
Weight of wire with end
   immersed in water . . . . = C

CALCULATIONS

$$\text{Specific Gravity} = \frac{B - A}{B - A + C - D}$$

You must recognize the possible presence of plasticizers, fillers, etc., when interpreting this measurement, for their presence can cause the specific gravity to differ greatly from that of the plastic itself. Specific gravities (at 73 deg. F.) of various plastics are listed in the reference chart, page 272.

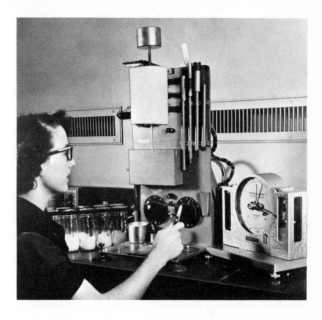

Fig. 4-25. Technician using test equipment.
(U.S. Industrial Chemicals)

Another method of determining the specific gravity of plastics materials is by the use of a direct reading specific gravity balance, Fig. 4-24. This balance is of special value in a polymer testing laboratory.

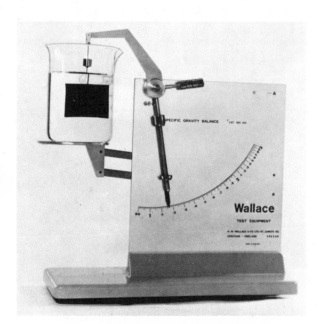

Fig. 4-24. This direct reading specific gravity balance eliminates time-consuming labor and calculations. The specific gravity of the plastic sample is read at the dial pointer.

Fig. 4-26. This is a melt indexer with digital temperature display, elapsed time indicator and weight support. The weight at the top cylinder provides the extruding force.
(Tinius Olsen Testing Machine Co.)

Fig. 4-27. A visual cube appearance test is being carried out on a clean table surface. Jars of color control samples are in the background.

Fig. 4-29. Hardness of a resin sample is measured with Shore durometer. This instrument indicates on a dial how far a steel indentor penetrates into the sample.

MELT INDEX

The melt index of a resin describes the flow behavior that can be expected during processing. It usually varies due to the average molecular weight of the polymer. The melt index refers to the amount of resin in grams of a thermoplastic material which can be forced through a 0.0825 in. orifice when subjected to 2160 grams of force in 10 minutes at 190 deg. C. If the melt index of a resin is high, the melt flow resistance during processing is low. That is, the material flows through a mold or extruder die faster than a resin with a low melt index. Fig. 4-25 shows a melt index test being performed on polyethylene; Fig. 4-26 illustrates a typical melt index test machine. In a more practical sense, the melt index measures the ability of a molten polymer to flow and fill out a mold cavity uniformly. If a given polymer used in an injection molding operation causes continuous mold flashing problems, the melt index of that polymer may be too high. The polymer is too thin (low viscosity) at that molding temperature. When this happens at given molding temperatures, the polymer flows too easily. It is not thick enough to flash the mold. A polymer with a lower melt index may solve the problem.

A high melt index number indicates that the resistance of a polymer to processing pressure and viscosity are low. Polymers have melt index numbers that range from 0.25 to more than 25. A melt index of 1.0, for example, would indicate that a polymer would be high viscosity (thick) at a given processing temperature.

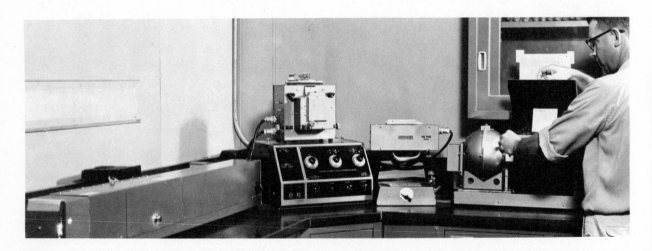

Fig. 4-28. This test equipment includes a color difference meter, a gloss meter, and a haze meter. They measure color uniformity, surface gloss, and film haze.

RESIN COLOR AND APPEARANCE

Control tests are made in the laboratory to check for lot-to-lot uniformity of natural granules, Fig. 4-27. This includes the natural whiteness of nonpigmented resin, uniformity of pellet size, and a check for contamination due to dust, lint, or other foreign matter.

Color uniformity in resins is important so product color may be maintained. Automobile taillights, for example, are required by law to meet exacting color tolerances. Extensive laboratory tests are carried out by producers of resins to insure the processor of dependable colored plastics, Fig. 4-28.

HARDNESS

Hardness is an important factor in the selection of a polymer for some products. High pressure laminates used for counter tops, for example, provide for better abrasion resistance and wear if the surface is hard.

The hardness of plastic materials is relative to many properties and does not define a single property. In general, the hardness refers to scratching, resistance to indentation, and abrasion. The test for hardness is usually restricted to the indentation of a round rod into the plastic surface and is measured by such devices as Rockwell hardness testers and the Shore durometer, Fig. 4-29. Comparison of hardness values can be located in the reference tables, page 277.

TENSILE STRENGTH

Tensile strength refers to the resistance of a plastic material to be pulled apart. As noted in the reference section, page 272, plastic materials range from 2,000 psi to 30,000 psi in tensile strength. However, this is not necessarily a true measure of their ability to withstand a pulling force. Some plastics, such as polypropylene, stretch many times their length before they rupture. In many cases the processor may require high elongation strength in the product.

Fig. 4-30. This is a standard tensile test specimen. It has been accurately produced to this shape by injection molding. (Union Carbide Corp.)

Tensile strength at rupture and elongation are the two properties of most importance.

Standard test samples, Fig. 4-30, are used to measure tensile strength. Some samples are injection molded while others are die cut from sheet stock, Fig. 4-31. A universal press can be used for stamping a variety of laboratory test specimens including those for tensile testing, Fig. 4-32.

A number of valuable properties can be determined by the use of a tensile testing

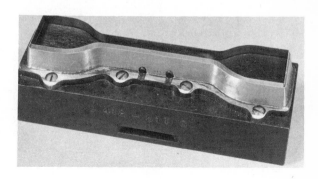

Fig. 4-31. A steel die used for stamping tensile test samples from sheet plastic material. (M.S. Instrument Co.)

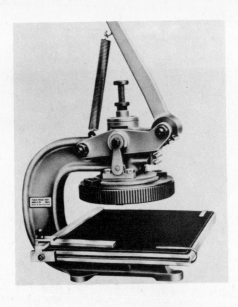

Fig. 4-32. This hand operated press can be fitted with a die to stamp tensile test specimens. (M.S. Instrument Co.)

machine, Fig. 4-33. Elongation before rupture of a test specimen, Fig. 4-34, can be measured while the test is in progress, Fig. 4-35. Another instrument, called an extensometer, may be used to determine the amount of strain under load on the plastic sample, Fig. 4-36.

CREEP RESISTANCE

Creep refers to the slow dimensional change of a plastic material when it is placed under load for a long period of time. In many cases creep can become a serious problem in fabricated products such as containers which must resist extensive loads for long periods of time without deforming.

Fig. 4-33. Automatic tensile testing machine with plastics specimen being clamped in place by technician. (Molded Fiber Glass Companies)

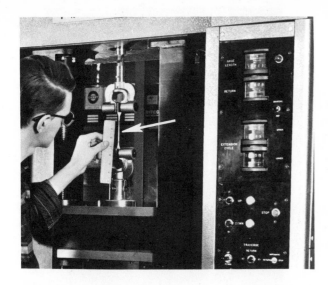

Fig. 4-34. In tensile test lower jaw of machine is pulled downward. A ruler is used to check amount of stretch. (U.S. Industrial Chemicals)

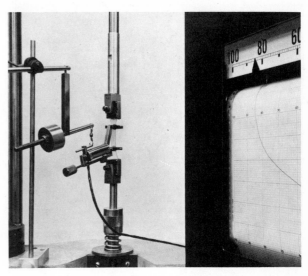

Fig. 4-36. This extensometer is graphically plotting a stress-strain curve on the chart at the right as the test is in progress. (Tinius Olsen Testing Machine Co.)

Temperature is a key factor in creep, as elevated temperatures usually increase this property. Creep at room temperature is known as cold flow and shows up in such resins as polyethylene under load. Creep is involved mainly with thermoplastic materials as most thermosets are quite rigid.

PERMEABILITY

The permeability of plastics is of special concern when dealing with film and sheet

Fig. 4-35. This close-up shows the elongation from the pulling force. (U.S. Industrial Chemicals)

material. Permeability means the ability of a plastic film to transmit water vapor or gasses. This property is important in the food packaging industry where fresh foods can be packaged in certain films that "breathe" and allow the foods to remain fresh longer. Polyvinyl chloride film allows oxygen to pass through and keeps meats and vegetables looking fresh. Cellulose acetate film has a high water vapor transmission rate making it suitable for many food packaging uses. Laboratory testing for permeability requires precision equipment, Fig. 4-37.

STIFFNESS AND FLEXIBILITY

Many plastics products require thin walls, stiffness and low cost. A blow molded gallon milk bottle requires a minimum of material and is easier to handle when made from a stiff resin like high density polyethylene. On the other hand, flexibility is of value in designing squeeze bottles for hair shampoo or plastic tubing which must be coiled.

Three types of stiffness are of particular concern; FLEXURAL STIFFNESS, TORSIONAL STIFFNESS, and FOLDING ENDURANCE. FLEXURAL STIFFNESS is calculated

Fig. 4-37. This water vapor transmission tester gives permeability reading on the scale to the right indicating how much vapor has passed through the film specimen.
(Honeywell, Inc.)

Fig. 4-39. This torsional stiffness tester applies a specified twisting force to the plastic test piece and records the force required.

by bending the plastic specimen a given amount and recording the load required to make the bend, Fig. 4-38. TORSIONAL STIFFNESS refers to the amount of twisting a plastic specimen can absorb, Fig. 4-39. It is measured by the force required to twist the sample a given amount. The sample is

immersed in a liquid and kept at a standard specified temperature. The FOLDING ENDURANCE test displays the flexibility of a plastic material as to its hinge properties. The knowledge of how many times a resin of given thickness can be flexed at a given point gives the designer an opportunity to provide for "built-in" hinges on many products, Fig. 4-40. The folding endurance test-

Fig. 4-38. A plastic sample of polyethylene is being deflected on this stiffness tester while the load required is recorded on the scale at the top.

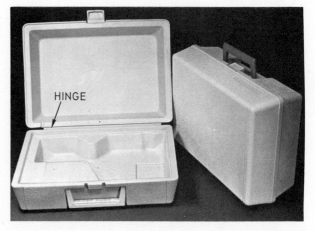

Fig. 4-40. Portable record player case made of high density polyethylene with integral hinge.

ing machine, Fig. 4-41, runs continuously until the test piece fails.

IMPACT STRENGTH

IMPACT STRENGTH is actually a measure of toughness of a plastic resin. Although toughness is not clearly defined, it is generally agreed that it is a measure of the energy required to break or rupture a plastic specimen. A standard measure of impact strength is illustrated in Fig. 4-42, where a pendulum strikes the test piece and the force at rupture recorded. Some plastics, such as polycarbonate and epoxies, approach the impact strength of steel, while many of the softer plastics like flexible polyurethane are difficult to measure since the test may

Fig. 4-42. The test piece is clamped in place in the vise of the impact tester and the pendulum is ready to be dropped. Impact force is registered on the scale at the top.
(Testing Machines, Inc.)

tear the material and not reveal its true strength.

Fig. 4-41. A folding endurance test in operation. The test piece is bent back and forth until failure while the number of times the piece is flexed is recorded on the dial.
(Tinius Olsen Testing Machine Co.)

Fig. 4-43. The impact strength of blow molded bottles of high density polyethylene is tested on this piece of equipment. Filled bottles are dropped from various heights to determine the height of burst.

Fig. 4-44. Setup used to determine the moisture content of resins. A radio frequency oven drives moisture from the resin into dry nitrogen. An electric cell determines the amount of moisture. (U.S. Industrial Chemicals)

In most industrial applications the impact strength is tested during product development. The results of these tests will often vary from those of standard specimen tests. A test of this nature is illustrated in Fig. 4-43. Similar tests for film materials are used to determine impact strength.

MOISTURE CONTENT

The absorption of moisture in resins is an important property that must be determined prior to processing. High moisture content can cause great difficulty in processing, such as blistering and surface marks, and also damage to machinery due to corrosion. Moisture content determinations indicate whether the resin should be dried before or during processing. Nylon, for example, due to its high rate of moisture absorption, must be kept dry prior to processing. Fig. 4-44 shows a moisture content test being performed on a resin sample.

The reference section, page 272, indicates the moisture absorption of many commercial resins.

HEAT RESISTANCE

As a group, plastics are poor conductors of heat and display good insulating properties. The THERMAL CONDUCTIVITY of plastics is measured by the volume of heat that will pass through a specimen in a given time for a given temperature. The unit of measure is the calorie; the volume of heat necessary to raise the temperature of one gram of water 1 deg. C. A comparison of thermal conductivity of plastics and aluminum indicates that aluminum will conduct heat between two and three thousand times as much as the average plastic material.

The resistance to heat is also indicated by classifying plastics as nonburning, self-extinguishing, or burning. Most of the thermoplastics do not meet the nonburning classification. Some thermoplastics burn with ease while others are self-extinguishing, Fig. 4-45. Certain fillers can be added to some thermoplastics to make them self-extinguishing.

Of particular importance to the designer is the heat distortion temperature and safe continuous operating temperature of plastics while in use. The reference section includes information on the use of many plastics at high temperature conditions.

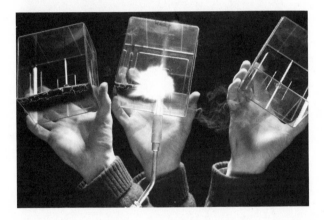

Fig. 4-45. The self-extinguishing properties of polycarbonate are shown here as the material burns while in the flame but does not support combustion as it leaves. (General Electric Co.)

ELECTRICAL PROPERTIES

The electrical properties of plastics are considered under the topics of DIELECTRIC CONSTANT, ARC-RESISTANCE, DIELECTRIC STRENGTH, and DISSIPATION FACTOR.

The DIELECTRIC CONSTANT is used as a measure of the plastic's ability to insulate when used as an element in a capacitor. A high dielectric constant indicates the plastic material is suitable for use as an insulator in separating the electrodes of a capacitor.

ARC-RESISTANCE refers to the ability of a plastic material to resist an electric arc which would tend to burn its way across or through the plastic when in use as an insulator. It is measured in the number of seconds it takes for a given electrical charge to burn its way across the insulator, charring it enough to make it conductable. Arc-resistance is of extreme importance in the use of plastics for switch plates, circuit breakers, and high voltage insulators.

The DIELECTRIC STRENGTH of a plastic material indicates its suitability as an electrical insulator. Most plastics are excellent insulators in this respect. The dielectric strength is measured by the maximum current that can be caused to flow through a film of plastic .001 in. in thickness. More and more uses of plastics are being made in this respect as a replacement for other materials, Fig. 4-46.

Fig. 4-46. Electrical pin insulator made of polycarbonate with a high dielectric strength used on telephone line. (H.K. Porter Co.)

The amount of power lost in a plastic insulator is known as the electrical DISSIPATION FACTOR. It is a measure of the percentage of electrical energy lost as heat within the plastic when used as an insulator. A low dissipation factor is an advantage in most electrical applications as it conserves energy and does not harm the plastic by overheating. An example of the advantage of a high dissipation factor is in heat sealing film. Cellulose acetate heat seals exceptionally well due to its high dissipation factor causing the material to melt and flow quickly. Polypropylene film, on the other hand, cannot be sealed electronically since it absorbs the electrical energy which would cause it to soften.

CHEMICAL RESISTANCE

The resistance of plastic materials to attack by chemicals must be considered for most every product. Each particular resin displays different reactions to the hundreds of chemicals with which it may come into contact. For specific applications, such as tubing and containers for chemical laboratories, tests must be evaluated for each chemical the plastic is expected to resist, Fig. 4-47.

Standardized chemical tests are carried out at 75 deg. F. for chemical resistance. Some of the effects of chemical testing with water, salts, solvents, and other chemicals may cause:

1. Change in color.
2. Surface deterioration.
3. Swelling or shrinkage.
4. Loss of stiffness or strength.

5. A change in weight due to absorption. General chemical resistance properties are listed under the chapter on individual polymers. Specific properties for any polymer can be obtained from the manufacturer's literature.

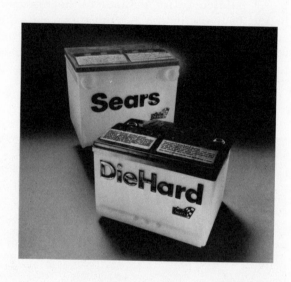

Fig. 4-47. Products like these batteries require the selection of polymers that will resist strong acids.
(Gladen Div., Hayes-Albion Corp.)

TEST YOUR KNOWLEDGE - CHAPTER 4

1. What two main safety precautions should be observed when running a burning test on a plastic sample?
   a._____.
   b._____.
2. The heated copper wire test is used to determine if_____is present in the unknown sample.
3. Bending a plastic test sample and measuring the force required to make the bend measures the property of_____.
4. The melt index of a plastic is measured in_____.
5. The electrical properties are divided into the following four categories.
   a._____.
   b._____.
   c._____.
   d._____.

6. What two methods are suggested for cutting test strips from sample sheet material? Why is each used?
   a._____.
   b._____.
7. Explain the test for determining whether a resin sample is thermoplastic or thermosetting.
8. What five characteristics should be observed when performing a burning test for plastics identification?
   a._____.
   b._____.
   c._____.
   d._____.
   e._____.
9. Moisture content of a polymer is expressed:
   a. In grams.

b. As a percentage.

c. As a ratio.

d. Any of the above.

10. Define tensile strength as it refers to plastics.

11. The three types of plastics stiffness generally tested are:

a. _____.

b. _____.

c. _____.

12. Impact strength in plastics is a measure of:

a. Toughness.

b. Hardness.

c. Rigidity.

d. Ductility.

e. None of the above.

13. The term _____ is used to indicate slow dimensional change in a plastic when it is under load for a long period of time.

14. The weight of a specimen of plastics compared to the weight of an equal volume of water is known as _____.

15. When preparing sheet material from granules, how can you determine when complete flow in the mold is obtained?

16. In what two ways are test specimens for tensile strength tests made?

a. _____.

b. _____.

17. Define permeability in reference to plastic film.

18. Chemical attack on plastics may cause any of the following conditions.

a. _____.

b. _____.

c. _____.

d. _____.

e. _____.

SUGGESTED ACTIVITIES

1. Design a sheet metal layout and construct the mold for forming sheet plastic from granules. Make the mold size suitable to the capacity of your laboratory hydraulic press.

2. Make sample plastic sheet material of resins suggested by your instructor. Cut the sheet into test strips and perform a burning identification test on each resin. Record your results and compare them with those on the burning chart. Check odors with known test tube samples.

3. Collect at least six different plastics products which can be tested. Cut a test strip from each product and, using the burning test, try to identify each resin. Compare your results with the burning chart.

4. Using the remaining test strips from your sample sheet, cut them to equal size, polish the edges, and mount them on a chart for resin classification. Include the polymer name, trade name, manufacturer, and color for each strip and indicate the outstanding properties of each.

5. Perform standardized test with plastic samples on each testing machine available in your laboratory. Record results and write a report on your findings.

6. Select an unknown sample of a polymer and, using your test equipment, determine the specific gravity of the specimen.

7. Prepare a bulletin board display depicting a number of plastics tests. Design the display around the properties resulting from the tests and their value illustrated in plastics products.

8. Visit a plastics company and find out what tests they are performing on plastics for the products they manufacture.

9. Present a demonstration on the proper way to set up the hydraulic press for sample molding.

10. Prepare a series of posters that deal with safety in making burning identification tests on plastics.

Chapter 5

INJECTION MOLDING

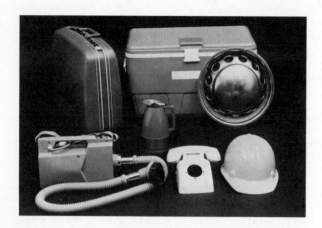

Fig. 5-1. Various types and sizes of injection molded parts include a vacuum cleaner housing, suitcase, ice chest, electroplated wheelcover, hard hat, telephone and pitcher. (Uniroyal, Inc.)

Many different processes are used to transform plastic granules, powders, and liquids into final products. The plastic material is in moldable form, and is adaptable to various forming methods. In most cases thermoplastic materials are suitable for certain processes while thermosetting materials require other methods of forming. This is recognized by the fact that thermoplastics are usually heated to a soft state and then reshaped before cooling. Thermosets, on the other hand, have not yet been polymerized before processing, and the chemical reaction takes place during the process, usually through heat, a catalyst, or pressure. It is important to remember this concept while studying the plastics manufacturing processes and the polymers used.

Fig. 5-2. A 32 ounce reciprocating screw injection molding machine with 375 ton mold clamping force. This is a rapid production machine. (Cincinnati Milacron Co.)

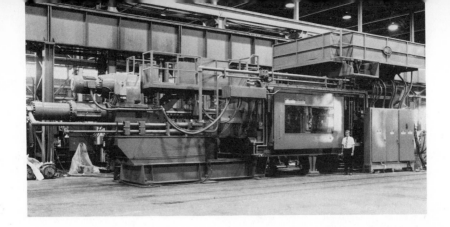

Fig. 5-3. A large reciprocating screw injection molding machine, with 320 ounce (9.01 kg) shot capacity and mold clamping force of 2,700 tons. (National Automatic Tool Co.)

INJECTION MOLDING

Injection molding is by far the most widely used process of forming thermoplastic materials. It is also one of the oldest. The basic process involves six major steps in the molding cycle:

1. The hopper is loaded with granular plastic materials.
2. Heat is applied to the plastic until it becomes soft enough to flow.
3. The softened plastic is forced through a nozzle into the mold cavity.
4. When cool, the halves of the mold are separated.
5. The part is ejected from the mold.
6. Gates connecting the product to the runner system are removed.

Variations of this basic molding process are involved in all injection molding. Fig. 5-1 illustrates some typical products produced by injection molding.

INJECTION MOLDING MACHINES

Injection molding machines are manufactured in many sizes. These are rated according to size by the amount of material which can be injected in one cycle, which ranges from a fraction of an ounce in the small laboratory models to many pounds in

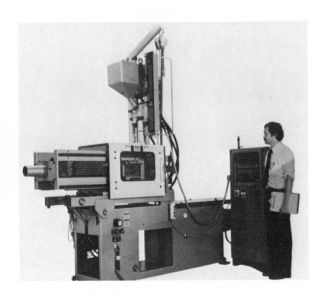

Fig. 5-4. This versatile laboratory (or production) 40 ton screw injection molding machine is equipped with a microprocessor control unit. (Polymer Machinery Corp.)

Fig. 5-5. Hydraulic reciprocating screw injection unit shows hopper, heating elements and nozzle. (National Automatic Tool Co.)

Fig. 5-6. Molding unit of an injection molding machine. Mold halves are mounted on movable and stationary platens. (Cincinnati Milacron Co.)

large production equipment, Figs. 5-2 and 5-3. Laboratory models, Fig. 5-4, are used in the research and development of new polymers and molding techniques.

There are two basic units to an injection molding machine; one for injecting the heated plastic and one for opening and closing the mold. The first unit, Fig. 5-5, includes a feed hopper, heated injection cylinder, and an injection plunger or screw system. The second unit comprises a hydraulic operated moving platen and a stationary platen on which the halves of the mold are mounted, Fig. 5-6. Injection molding machines are also available in vertical models, Fig. 5-7.

There are many variations in injection molding machine design, however, the basic machines are of either the screw-ram, Fig. 5-8, or plunger type, Fig. 5-9. The main difference between these types is the method in which the plastic material is delivered from the hopper to the nozzle of the machine. Machines of the reciprocating-screw type are used more because of faster cycles, lower melting temperatures, and better mixing of the material. Our description of injection molding will involve this type of screw-ram machine, Fig. 5-10.

Fig. 5-7. Vertical injection molding machines are often used for producing small, close tolerance parts. This machine is a four ounce reciprocating screw type.

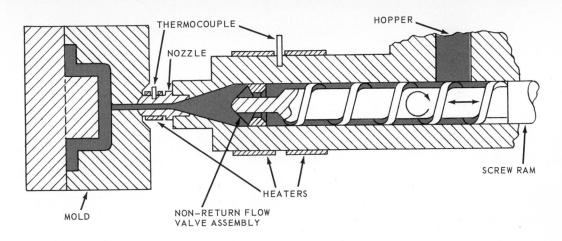

Fig. 5-8. Plastic material flow from the hopper to the mold on a reciprocating injection molding machine.
(Eastman Chemical Products, Inc.)

MOLDING PROCESS

Plastic granules are fed into the hopper and through an opening in the injection cylinder where they are carried forward by the rotating screw. The rotation of the screw forces the granules under high pressure against the heated walls of the cylinder causing them to melt. As the pressure builds up, the rotating screw is forced backward until enough plastic has accumulated to make the shot. At this point the screw is hydraulically forced forward, injecting the plastic through the nozzle, on through the sprue and runners, and into the cavities of the closed mold. Pressure is held briefly in order for the plastic to set, after which the screw retracts, releasing the pressure. Water cooled molds cause the plastic to cool quickly. The mold is opened and the part is ejected from the movable half of the mold, usually by air pressure or spring loaded ejector pins. The mold is then closed to begin another cycle. The complete process is controlled by timers which are set to break the cycle down into each function. Newer machines incorporate microprocessor units to automatically control the entire process.

INJECTION MOLDS

Molds used in injection molding consist of two halves; one stationary and one movable. The stationary half is fastened directly

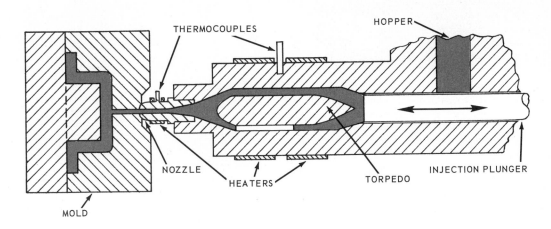

Fig. 5-9. Plastic material flow from the hopper to the mold on a plunger type injection molding machine.
The torpedo or spreader forces the plastic against the cylinder walls for even heating.

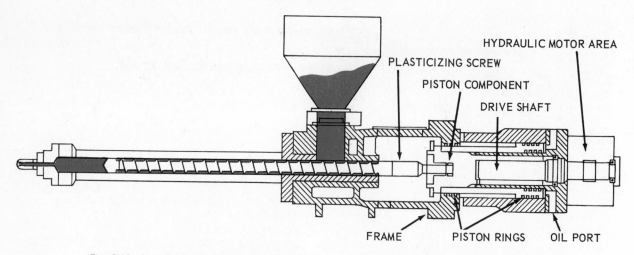

Fig. 5-10. A reciprocating screw machine in the back position. The screw is ready to be forced forward to make the injection shot. (Cincinnati Milacron Co.)

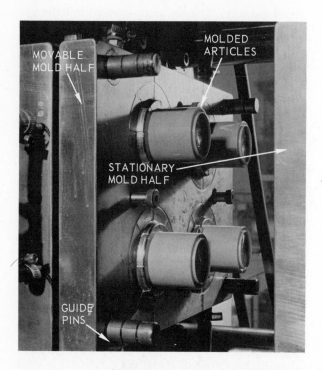

Fig. 5-11. The movable half of a mold in the open position. Molded articles have been pulled from the sprue and are ready to be ejected by air pressure.
(Amoco Chemicals Corp.)

Fig. 5-11. There are many possible mold designs, including multiple piece molds for complicated parts. A typical two-piece mold is shown in Fig. 5-12. On production injection molding equipment many articles may be shot at the same time by the use of multiple cavity molds. The use of a balanced

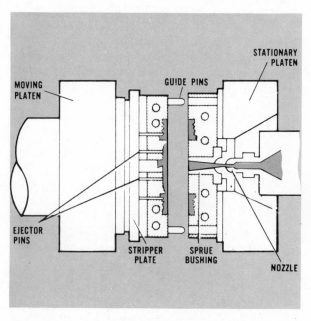

Fig. 5-12. A typical two-piece, two cavity injection mold. This shot includes the sprue, runners, gates, and products. Water cooling maintains mold temperature.

to the stationary platen and is in direct contact with the nozzle of the injection unit during operation. The movable half of the mold is secured to the movable platen and usually contains the ejector mechanism,

Fig. 5-12A. These are eight track tape cases in a multiple cavity mold. The runner has been severed from the parts prior to ejection. (Alma Plastics Co.)

vide for even filling of mold cavity and to allow products to be easily removed from the runner system, Fig. 5-12A With most injection molding systems, the articles can be snapped away from the runner or sprue without additional trimming.

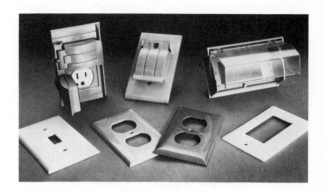

Fig. 5-14. Polycarbonate switch and outlet cover plates are made by the million using injection molding. (Society of the Plastics Industry)

runner system carries plastic from sprue to each individual cavity. At this point, material passes through a gate into cavity. The gate is a restriction, smaller than the runner, to pro-

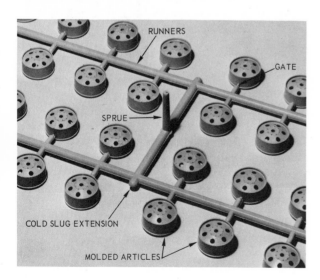

Fig. 5-13. Molded articles will be trimmed at gates. Remaining sprue and runner system will be reground and returned to hopper. (Mobay Chemical Corp.)

Fig. 5-15. A molded polystyrene cabinet door being removed from a production injection molding machine. (Shell Chemical Co.)

Fig. 5-13 shows the product from an injection mold before trimming. Products that have been injection molded can usually be identified by finding where the gate was broken off. The gate will usually be located at the edge or parting line of an object or in the center of cylindrical products.

Injection molding is especially suitable for high production runs, Fig. 5-14. Molds are expensive, as are the machines. Yet, once the product has been designed, molds made, and production started, articles can be produced in quantity at low cost, Fig. 5-15. Most machines produce several thousand products an hour. Virtually all thermoplastics can be injection molded through variations in mold and machine design. Fig. 5-16 illustrates the versatility in production of a typical injection molded product.

Fig. 5-16A. Runners are being reground in an auger granulator. (Polymar Machinery Corp.)

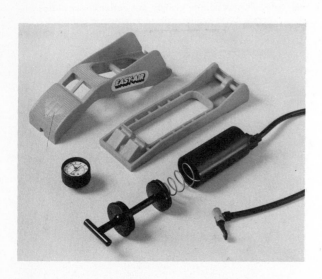

Fig. 5-16. Tough polypropylene resin is used to injection mold the components of this automotive tire foot pump. (ICI Americas Inc.)

AUXILIARY EQUIPMENT

In-plant resin coloring for injection and other molding processes is accomplished by mixing powder pigments or color concentrates with natural color granules. Final pigment concentration varies between one and five percent. Mixing is done in drum tumblers and takes up to one half hour.

Some thermoplastic materials contain so much moisture that they may blister when processed. To prevent blistering, they may be dried in an oven at temperatures just below their melting point. At this temperature, the moisture will be driven out quickly. High production molding processes often utilize heating driers contained in the machine hopper.

Mold chilling or heating equipment are needed to maintain the temperature of the mold. Water or antifreeze is pumped through channels in the mold to transfer away heat absorbed from the plastic. Some high temperature plastics often cool best at temperatures approaching 200 deg. F. (93 deg. C). Hot water or oil must then be used.

Injection molding operations generate scrap material from sprues, runners or reject parts. Since they are thermoplastic, they may be reprocessed. To return the material to a form which will allow it to be fed through a hopper, it must be ground up. Auger type granulators, Fig. 5-16A, receive runners directly from the machine where they are fed through cutting knives. The granulated plastic is vacuumed back to the hopper where it is mixed with new (virgin) material. Granulators placed beside the injection molding machines are hand fed by an operator. Contamination of the material with dirt or other plastics is the most troublesome problem of regrinding.

THERMOSET INJECTION MOLDING

A vast majority of injection molding makes use of thermoplastic materials. However, injection molding of thermosetting materials is done by using modified screw molding machines. Thermosets must be polymerized by heat to harden, instead of solidifying, by cooling. This is done by preheating the material in the cylinder to a plasticizing (softening) temperature of 150 to 240 deg. F. (65 to 115 deg. C). It is then injected into a hot mold and cured at 325 to 400 deg. F. (162 to 204 deg. C). After curing, the parts are ejected from the mold while still hot. Should the thermosetting material stay in the cylinder too long or if it is too hot, it will set up and clog the machine.

The cylinder of these special machines is heated by an oil or water jacket. Electrically heated cartridges inserted in the mold provide the product curing temperature, Fig. 5-16B.

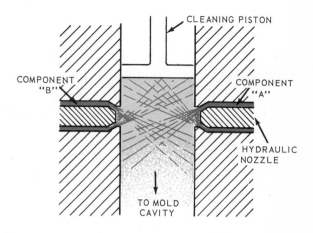

Fig. 5-17. This shows high pressure mixing of liquid components during a reaction injection molding cycle.

Phenolic, urea and melamine are the most common thermosets processed by screw injection molding. The major advantages of the process are the elimination of material preforming and preheating and a great reduction in cycle time as compared to more common compression and transfer molding processes.

It should be remembered that thermosetting scrap and defective parts CANNOT be reground and returned to the hopper for reprocessing.

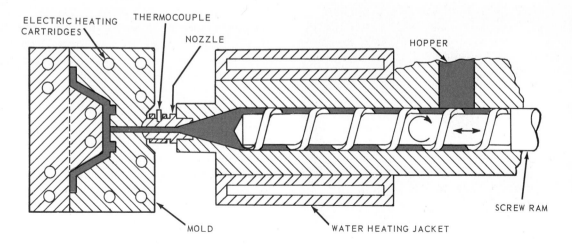

Fig. 5-16B. Thermoset injection molding using a modified reciprocating injection molding machine. Note water jacket and straight screw.

REACTION INJECTION MOLDING

The reaction injection molding process (RIM) involves the injection of a two component reactive polyurethane liquid into a closed mold. Two liquid streams are injected together under high pressure (2,500 psi or 17 237 kPa) into a mixhead, Fig. 5-17. They then flow through a gate and into the bottom of the mold. The air within the mold is allowed to escape through the parting line as the liquid rises and fills the cavity. The thermosetting urethane goes through a gel stage before curing. When the part has sufficiently cured, it is removed by hand, Fig. 5-18, or knocked out with ejector pins. Cycle times average between two and four minutes. Excess material (flash) commonly squeezes out of the mold at the parting line and must be trimmed off of the final product.

Mold release is sprayed on the cavity walls before each molding cycle to facilitate part removal. Mold release is slippery and must be cleansed from the surface of the part if it is to be painted. RIM parts are commonly painted. Examples include exterior car components which must match the color of the rest of the car.

RIM molding has the potential to become one of the largest processing techniques in the plastics industry. For large scale production, RIM has cost advantages over other plastics processes and metal fabrication. Total energy consumption and machine costs are lower than injection molding. Part size capacity is greater and parts may be integrated into one large complex product.

INJECTION MOLDING IN THE LABORATORY

The injection molding process can be studied through many forms of laboratory activity. The equipment need not be of industrial nature since the process involves the same fundamentals as industrial molding. In most cases a good understanding of the injection molding process can be obtained through the study of the equipment available and its capacity, the design of a suitable product, selection of a resin, and construction of the mold.

INJECTION MOLDING MACHINE

A typical laboratory injection molding machine, Fig. 5-19, will vary in shot capacity from a few grams to about one-half ounce. It will have a clamping device that is usually hand operated with a cam action and an injection system operated manually or by compressed air. The injection system exerts very high pressure on the mold, from 5,000 to 20,000 psi. It is therefore necessary for the mold to be clamped together with a greater pressure than the injection pressure or the mold halves will be forced apart and flash (squeeze out of resin at the parting line of the mold) will result.

The loading chute and the heating cylinder make up the other major components of the machine. Fig. 5-20 illustrates the detailed parts on the inside of an injection molding machine. Most of the laboratory size machines are of the plunger type.

PRODUCT DESIGN

A complete drawing of the product should be made to facilitate mold construction, Fig. 5-21. This should include the details

Fig. 5-18. Shell of this golf cart is molded in five parts from polyurethane elastomer by the reaction injection molding (RIM) process. (Union Carbide Corp.)

Fig. 5-19. This automatic plunger type injection molding machine is especially suited for laboratory study. (Kissam Manufacturing, Inc.)

of the product itself as well as the location of the sprue and the runner system with gates as required. Most thermoplastics are easily removed from the mold but it is usually advisable to place two degrees of draft on those surfaces perpendicular to the parting line. If venting is necessary, to allow for the escape of air compressed in the cavity, this should be noted on the drawing.

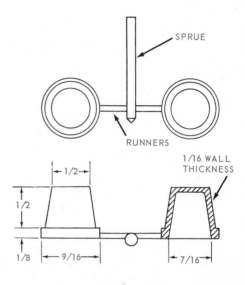

Fig. 5-21. Working drawing of salt and pepper shaker stoppers. Here a twin cavity mold with runners and sprue location provide the necessary details for mold construction.

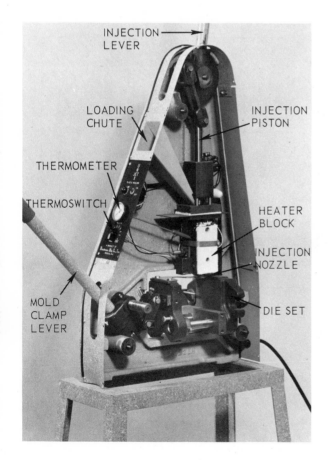

Fig. 5-20. This cutaway shows the working parts of a hand operated one-third ounce plunger type injection molding machine.

The required machining or casting equipment should also be taken into consideration during product design in order to determine whether it is feasible to make the mold. Fig. 5-22 shows some typical injection molded products produced on a one-third ounce machine. Figs. 5-23 and 5-24 give details of product and mold design possibilities.

SELECTION OF RESIN

Virtually all thermoplastic resins can be injection molded, however, some are more difficult to process and require specialized equipment. From a knowledge of polymer structure, it becomes evident that resins with a wide melting range mold

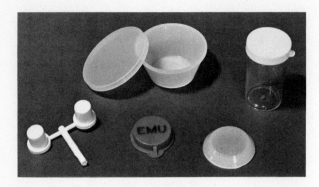

Fig. 5-22. Products made on a laboratory injection molding machine include salt and pepper plugs, soft drink bottle closures, and small parts containers with snap-caps.

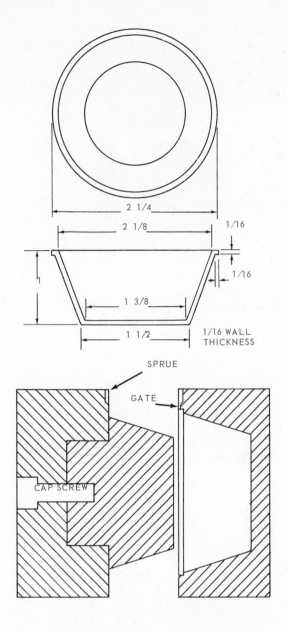

Fig. 5-23. Above. Design details of a small plastic parts container. Below. Suggested layout of the steel mold.

easier than those with a sharp or narrow melting point. Other factors to consider are the molding temperature requirements, moisture absorption of the resin, and the necessary properties required for the product. For example, acetal resin has a sharp melting point, turning quickly to a liquid, which requires exacting temperature controls. Polycarbonate requires high injection temperatures, while nylon absorbs moisture and will corrode the internal parts of the injection cylinder unless they are chrome plated. Product requirements may include such factors as elasticity necessary for a container snap-cap, clarity for a container, flex properties for a hinged part, toughness for a screwdriver handle, or hardness for ball-point pen.

The more common resins for ease of processing on laboratory injection molding equipment are the polyethylenes, polystyrene, polypropylene, styrene acrylonitrile, ABS, ionomer, and polyallomer. This group of resins provides ample selection for most injection molded product requirements.

MOLD CONSTRUCTION

The construction of the mold for injection molding begins with the working drawing. From it, the requirements for the mold can be specified. These would include the material from which the mold should be made,

the availability of equipment for machining the mold, and the mold capacity of the die set on the machine.

Cold rolled steel is an ideal material for laboratory molds, since it machines well, is fairly inexpensive, and holds up well for nozzle pressure and wear. Its major disadvantage is that it will rust quickly unless protected by mold release or wax during storage. Complicated mold cavities need specialized machining and polishing, therefore, circular cavities which can be turned and polished on the lathe require less equipment and machining skill.

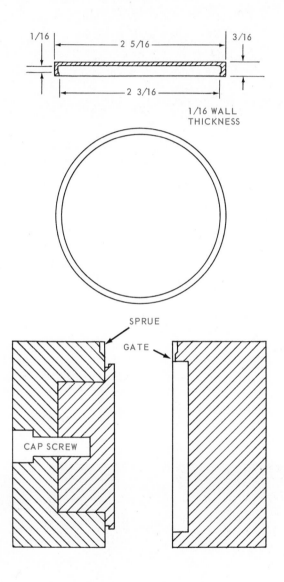

Fig. 5-24. Above. Drawing which shows details of the snap-cap for parts container. Below. Mold layout.

Fig. 5-25 illustrates a typical drawing of a product to be injection molded. Fig. 5-26 shows details of the mold. An alternative design is shown in Fig. 5-27. Construction procedure is as follows:

1. Secure two blocks of cold rolled steel

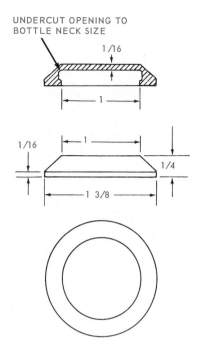

Fig. 5-25. Detailed drawing of a soft drink bottle closure. The neck of a bottle should be carefully measured for dimensions of the inside of the cap.

meeting the specifications of the machine mold capacity.

2. Drill dowel pin holes through one mold half and part way into the second half.

3. Press fit dowel pins into second half.

4. Fit mold half with dowel pins in four-

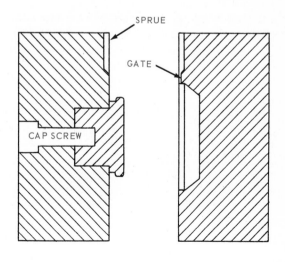

Fig. 5-26. A cross section of the bottle closure mold indicates construction details.

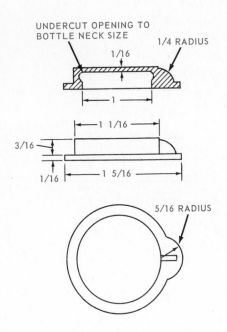

Fig. 5-27. This bottle closure design features a tab for easy opening and a rib for strength.

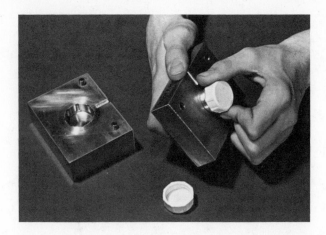

Fig. 5-29. An aluminum injection mold for a threaded bottle cap. After the mold is shot, the halves are separated and the cap unscrewed from the threaded plug.

jaw chuck of lathe and center to the middle of the cap.

5. Turn mold cavity as shown in drawing and polish.

6. Insert plug half of mold in four-jaw chuck, center, and bore holes for plug and cap screw.

7. Select stock for plug, place in three-jaw chuck, and turn to dimensions. Polish.

8. Insert plug in mold and draw down with cap screw.

9. Clamp mold halves together and drill

sprue hole to 1/8 in. from cavity.

10. Separate mold and file a half-round gate from end of sprue to cavity. Mold is then ready for use, Fig. 5-28.

Similar molds may also be machined from aluminum, Fig. 5-29, these have the advantage of not rusting. Excessive wear develops on the sprue due to the high nozzle pressure on the soft aluminum, but this can be overcome by the use of a steel cover plate on the top of the mold.

Fig. 5-28. The completed steel bottle closure mold illustrates the machining details.

Fig. 5-30. This aluminum filled epoxy mold was cast in a steel flask using two golf tee patterns. The steel top plate is attached with cap screws.

Another method of mold construction is by the casting process using an aluminum filled epoxy resin. This type of mold is particularly suited to products of intricate de-

Fig. 5-33. Mold is placed in the die set to the stop block and is locked in place.

Fig. 5-31. Set of interchangeable injection cylinders provides for quick change in resin and color. Cylinder to be used is lifted from rack.

sign and products that are difficult to machine. The cast epoxy is strong and gives good surface detail, however, it is brittle and should have a steel top plate attached to absorb the wear of the nozzle, Fig. 5-30.

mold has been cured, the cope is placed over it and the remainder of the mold poured. Upon curing, the flask is removed, all surfaces machined smooth, dowel pin holes drilled, and dowels inserted. A steel

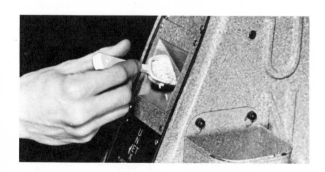

Fig. 5-32. Granules of the plastic are fed into the loading chute with a measuring spoon.

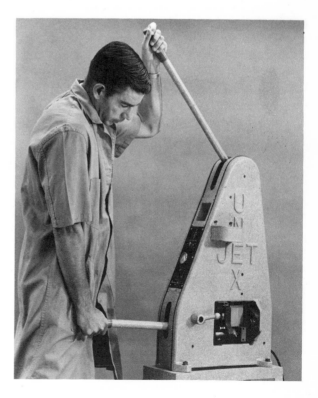

A pattern of the product must be secured or made and placed on a mold plate. The drag of a small steel flask is placed around the pattern and the epoxy resin is poured to fill the mold half. When this half of the

Fig. 5-34. Making the shot with the injection lever.

cap plate should be bolted to the top halves and the sprue, runners, and gates machined. Instructions for mixing, pouring, and curing the aluminum filled epoxy should be followed according to the manufacturer's specifications.

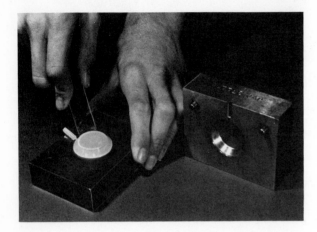

Fig. 5-35. Since the plug of the mold is undercut, the bottle closure must be pried off with a wedge stick.

MOLDING

Injection molding on a laboratory machine requires but a few basic steps. When the mold is completed, the desired resin for the product should be selected, Fig. 5-31, and the container placed near the loading

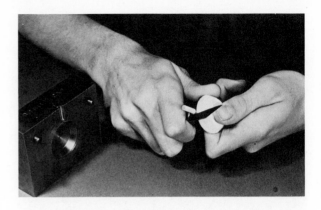

Fig. 5-37. Using sharp knife, trim the sprue from the closure at the gate.

chute of the machine. The temperature control should be set according to specifications for the resin. The granular resin is loaded into the chute, Fig. 5-32, to just below the top of the injection cylinder. The mold is inserted into the die set, Fig. 5-33, and locked in position. Injection of the shot is made and pressure is maintained on the injection lever for a few seconds to allow for cooling, Fig. 5-34. The mold should then be removed, the two halves separated, and the product ejected, Fig. 5-35 and 5-36. A wedge stick is recommended for removing many products which have to be pried from the mold. After removal, the sprue is detached, Fig. 5-37, and the closure is tested by snapping it on a bottle, half filled with water, to see if it is airtight, Fig. 5-38.

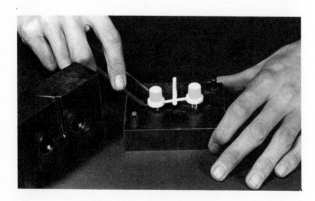

Fig. 5-36. A vacuum will often form around a tapered plug so these salt and pepper stoppers are being removed with a wedge stick.

Fig. 5-38. The closure is snapped on a bottle to check for fit and airtightness.

TEST YOUR KNOWLEDGE - CHAPTER 5

1. Injection molding machines are rated according to size by _____.
2. The six major steps in the injection molding cycle are:
   a. _____.
   b. _____.
   c. _____.
   d. _____.
   e. _____.
   f. _____.
3. The two basic units of an injection molding machine are:
   a. _____.
   b. _____.
4. The two main types of injection molding machines are the _____ or _____ type.
5. The molds on an industrial injection molding machine are opened and closed by _____ pressure.
6. The _____ on a product is usually located at the edge or parting line, or in the center of a cylindrical product.
7. Many products are molded at the same time by using:
   a. A higher speed.
   b. Many sprues.
   c. A runner system.
8. The opening in the mold where the product is formed is called the _____.
9. The easiest resins to injection mold:
   a. Have a sharp melting point.
   b. Have a wide softening range.
10. _____ is provided in a mold to allow for the escape of air.
11. Small injection molds may be cast from _____.
12. Forming thermoplastic materials by injection molding is:
    a. One of the most difficult methods.
    b. One of the most widely used methods.
    c. One of the newest methods.
    d. None of the above.
13. Why are thermosetting resins not used in standard injection molding?
14. Loading of an injection molding machine is done through the _____.
15. Parts are ejected from the mold by _____ or _____.
16. _____ of the plastic material is the most common problem associated with regrinding thermoplastic scrap.
17. When injection molding thermosetting plastics, you should normally heat the mold by _____.

SUGGESTED ACTIVITIES

1. Use molds available in the laboratory and injection mold products following the procedure for operating the machine.
2. Make a collection of products that have been injection molded commercially. For each product, describe or sketch how the mold was made. Look for the gate and marks left by the ejector pins.
3. Design a product and make a working drawing for a mold. Include a cross section of mold details. If materials are available, machine or cast the mold as required and make a trial product run.
4. Write to manufacturers of injection molding equipment asking for specifications on their machines. Write a report on the newest developments in injection molding technology.
5. Develop a chart illustrating ten resins being used in injection molding. Show the major products being made from these resins.
6. Make a display for the bulletin board of injection molded products. For each product indicate how it was "shot" and why it was made from a particular resin.

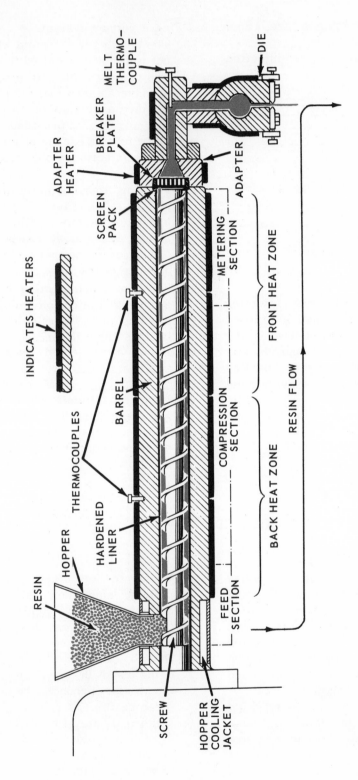

Fig. 6-1. Cross section of a typical extruder which shows the flow of plastic from the hopper to the die. The die will turn down as shown, or extend straight out from the extruder screw. (U.S. Industrial Chemicals)

Chapter 6
EXTRUSION

In the EXTRUSION process, thermoplastic resin is fed into a heated cylinder. The softened plastic is forced by a rotating screw, or plunger, through openings in accurately machined dies to form continuous shapes. See Fig. 6-1.

Extrusion is used to make three main types of products or for compounding:

1. Standard shapes such as rod, pipe, sheet, and irregular cross sections.
2. Film in single or multiple layers to be used alone or as coatings for paper, cloth and other surfaces.
3. Extrusions around wire and cable as a protective coating.
4. Compounding or mixing of additives into plastics (non-product application).

Each type can be made in varying sizes depending upon machine size. Extruders are sized by the diameter of the rotating screw. They range from large industrial extruders with screw diameters from 2 to 8 in., Fig. 6-2, down to the laboratory model with a 3/4 in. diameter screw, Fig. 6-3. Some typical extruded shapes are shown in Fig. 6-4 and 6-4A. Extrusion is used only for the processing of thermoplastic resins.

THE EXTRUSION PROCESS

Plastic granules are fed into the hopper of the extruder, picked up by the rotating screw within the hard liner of the extruder cylinder, Fig. 6-5, and are forced forward.

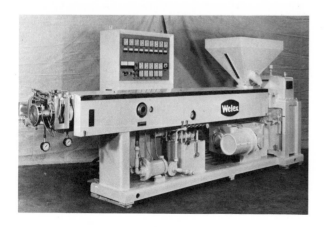

Fig. 6-2. This 4 1/2 in. (11.43 cm) extruder has a maximum output of 2,000 pounds (906 kg) per hour.   (Welex, Inc.)

Fig. 6-3. Sample testing, die development, and color matching are done on this 3/4 in. laboratory type extruder. (Wayne Machine and Die Co.)

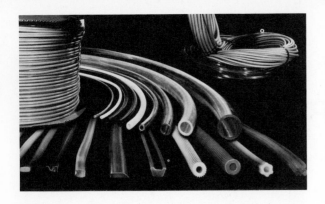

Fig. 6-4. These extruded products show a few of the many shapes that can be obtained by this process. (Stevens Elastomeric and Plastic Products, Inc.)

Fig. 6-4A. Tough packaging tape is extruded from polyvinyl chloride resin.

As the material moves along the cylinder, it is heated to a soft state and thoroughly mixed. The heat for softening the plastic comes from two sources: the external heater bands and the friction of the material against the rotating screw.

During the heating and compressing period the plastic must be transformed into a complete homogeneous (consistent) mix. This eliminates the possibility of wavy surfaces and a nonuniform cross section of the final product. It also provides for uniform color when the extruder is being used for color blending.

The melted plastic finally passes through the screen pack, which removes dirt, and to the die at very high pressure. From the die, the extruded profile passes through a cooling and take-off system, Figs. 6-6 and 6-7. The final extrudate is wound onto coils or cut to specified lengths.

FLAT SHEET AND FILM EXTRUSION

Flat sheet material is commonly extruded in thickness up to 1/4 in. (6.35 mm). Stock measuring under .010 in. (.254 mm) thickness is classified as film. The extruding sys-

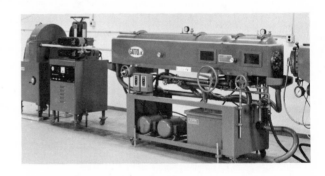

Fig. 6-6. An extrusion take-off system for pipe. Here the pipe enters the water cooling chamber from the die, is pulled out by the take-off belts, and is delivered to the cutting unit to be slit into specified lengths. (Gatto Machinery Development Corp.)

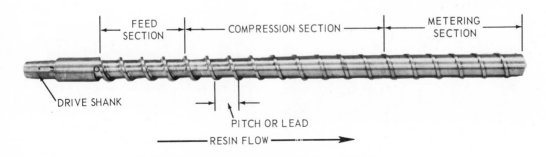

FEED SECTION — COMPRESSION SECTION — METERING SECTION

DRIVE SHANK

PITCH OR LEAD

RESIN FLOW

Fig. 6-5. The parts of a 3 1/2 in. extruder screw.

Fig. 6-7. This is a flat film extrusion take-off system. The extruded film passes through the polished roll stand, over the conveyor, through the pull rolls, and onto the winder. (Welex, Inc.)

Fig. 6-9. A 72 in. wide sheet extrusion die ready to be attached to the extruder head. (Waldron-Hartig Div.)

tem is the same used for other profiles, such as pipe and film, except for the die and the take-off equipment. See Figs. 6-8 and 6-9. Granular plastic is fed into the hopper of the extruder, Fig. 6-10, and goes through the conventional extruding system. At the die, Fig.

6-11, the soft, hot plastic is extruded directly into the finishing rolls at accurately controlled temperature. A conveyor system, Fig. 6-12, carries the sheet to the pull rolls, Fig. 6-13, where it is fed to the cutting unit. The sheet is cut to specified lengths ready for packaging or carried directly to thermoforming units for further processing. Quality control and inspection are required to assure proper dimensional tolerances, Fig. 6-14.

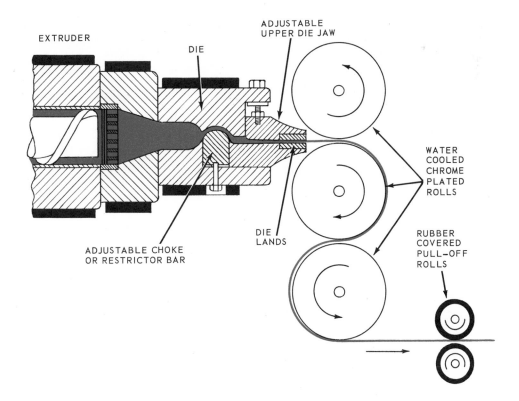

Fig. 6-8. Cross section of a die for sheet extrusion and part of the take-off unit. (U.S. Industrial Chemicals)

Fig. 6-10. High impact polystyrene being loaded into the hopper for a sheet extrusion test run.

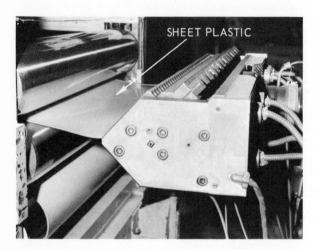

SHEET PLASTIC

Fig. 6-11. Sheet plastic flows from the extruder die to the polished finishing rolls of the take-off unit.
(Fellows Gear Shaper Co.)

Fig. 6-12. Sheet plastic passing from the cooling and polishing rolls onto the conveyor line.

FRICTION PULL ROLLS

Fig. 6-13. Friction pull rolls carry the sheet plastic from the conveyor to the traveling power shear to be cut to length and stacked.

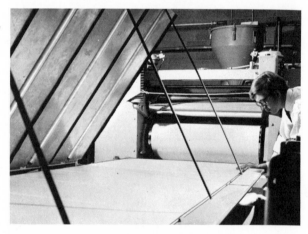

Fig. 6-14. Inspection of extruded sheet for quality control.
(Amoco Chemicals Corp.)

Extrusion coating is accomplished by pressing a hot plastic film extrudate and a substrate material between pressure rollers, Fig. 6-15. No adhesives are needed as the hot plastic bonds readily with the substrate. Different substrates include paper, cloth and metal foil. Standard film extruders are used for this process.

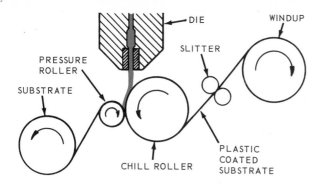

Fig. 6-15. Note diagram of the extrusion coating process.

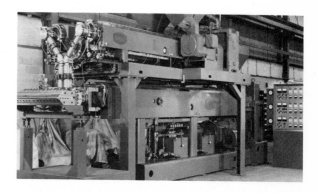

Fig. 6-15C. This is a film coextrusion machine with three parallel extruders individually controlled from one console. (Welex, Inc.)

Multilayer plastic films are produced by a coextrusion process, Fig. 6-15A. Two or more extruders are used to feed different plastics (or colors) into one die, Figs. 6-15B

and 6-15C, by means of a manifold system. These special films have properties utilizing characteristics of each plastic. Most are used for food packaging as barriers to moisture and gas vapors. Coextrusion is also used for other products such as two colored drinking straws or automotive windshield trim.

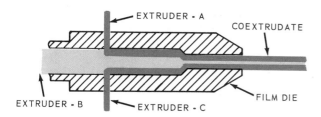

Fig. 6-15A. Feeding three different plastics materials through a film die, this process is known as coextrusion.

Fig. 6-15B. Two extruder barrels feed a film die in this overhead extruder. (Welex, Inc.)

Fig. 6-15D. A blown film bubble is being cooled by air ring as it rises up through sizing orifice. (Alpine American Corp.)

BLOWN FILM EXTRUSION

Blown film extrusion is similar to the flat extrusion process except the die forms a hollow tube, Fig. 6-15D, through which air is forced to expand the film into a cylinder, called a "bubble," Fig. 6-16. As the plastic bubble solidifies, it is squeezed together between rolls to form a double thickness film. It has been found satisfactory to extrude and pull the bubble upward, flattening it between rollers, and carrying it to a windup roll, Figs. 6-17 and 6-17A. In some cases, the film is cut to short lengths. It is then sealed at one end to form plastic bags. The film can also be split lengthwise and used for larger sheeting.

WIRE COATING EXTRUSION

The plastic coating of wire is another major use of the extrusion process. It is similar to the extrusion of pipe or tubing except the mandrel in the die is replaced by a tapered guide through which a continuous line of wire is fed, Fig. 6-18. As the plastic flows from the extruder through the die, it surrounds the moving wire which is preheated to the plastic melt temperature and leaves the die as a integral unit.

Many thermoplastic resins are used in wire and cable coating. The polyethylenes, polyvinyl chloride, and nylon are typical wire coating resins. Silicon is often used

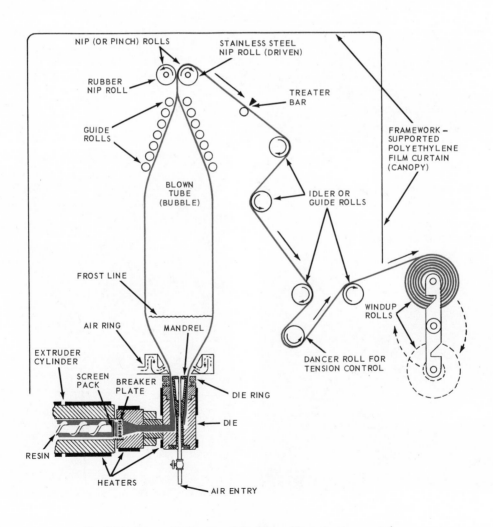

Fig. 6-16. Schematic drawing of the blown film extrusion procedure.
(U.S. Industrial Chemicals)

Fig. 6-17. Study blown film lines, folded low density polyethylene film on left and high density polyethylene film on right. (Hoelscher Corp.)

for high heat resistant applications. Fig. 6-19 shows the complete line in a wire coating system. Fig. 6-19A illustrates a few examples of coated cable.

EXTRUSION FOR COMPOUNDING AND GRANULATING

In a resin compounding plant, the extruder is used for the blending, coloring, and granulating of resins to be shipped to the processor. The plastic resins often require additives for particular applications along with the required colors desired by the processor. These are all added to the resin and fed into the extruder to provide a homogeneous mix in the resulting granules. A special granulate die is used on the extruder, Fig. 6-20, which produces many strands of resin. These strands are then cut into small pieces approximately 1/8 in. in length and pulled into loading tanks through an exhaust system, Fig. 6-21. The granulated resin is normally packaged in fifty pound bags or one-thousand pound cartons ready for shipment. Typical granulated resin is in the form of round, cylindrical, or cube shaped particles, Fig. 6-22.

Resin coloring is done either by adding dry powder color during the compounding

Fig. 6-17A. Blown film is continuously wound onto take-off roll. (Rextrusion Systems, Inc.)

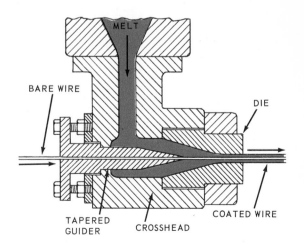

Fig. 6-18. A crosshead holds the wire-coating die and the tapered guider as the soft plastic flows around the moving wire.

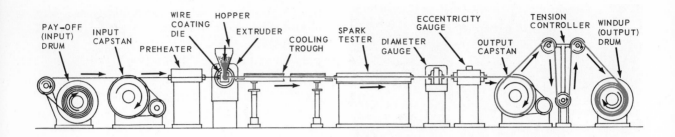

Fig. 6-19. A general layout of the components in a wire-coating extrusion plant.
(U.S. Industrial Chemicals)

process or by the use of color concentrates at the time of molding. Color concentrates appear as small granules similar to the resin except that they are almost pure color pigment bonded together by a small amount of the same resin used in molding. The normal coloring ratio is 5 lbs. of color concentrate to one-hundred pounds of natural resin.

Regrinding of thermoplastic materials from the extrusion process, as well as other molding processes, makes use of scrap and

Fig. 6-20. This hot melt granulator shows the many strands of plastic being extruded through the die into the grinder below. (Fellows Gear Shaper Co.)

Fig. 6-19A. Various sizes of wire cable coated with polyvinyl chloride, PVC is used because of its flexibility and abrasion resistance. (The B.F. Goodrich Co.)

Fig. 6-21. Drums are loaded with resin from the granulating system. (Farrel Connecticut Div.)

Fig. 6-22. Plastic granules from extrusion process ready for molding.

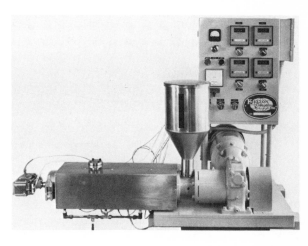

Fig. 6-24. This small 3/4 in. bench model extruder can be used for laboratory testing, color compounding, and trial production runs.   (Killion Extruders, Inc.)

reject resin. Trimmings from extrusions may be reground, Fig. 6-23, and remolded. The reground material is usually mixed with new material of the same grade and color.

EXTRUSION IN THE LABORATORY

The extrusion process can be performed in the plastic laboratory as a thorough study of this phase of the industry. A number of laboratory extruders are available, as in Fig. 6-24, which operate on the same basis as the large industrial machines. Dies may be purchased for the extrusion of round rod, Fig. 6-25, flat tape, and other profiles in various sizes. Dies may also be machined

Fig. 6-23. An industrial regrinder, this machine may be used separately or connected for automatic return of the reground resin, by vacuum, to the hopper of the extruder. (Foremost Machine Builders, Inc.)

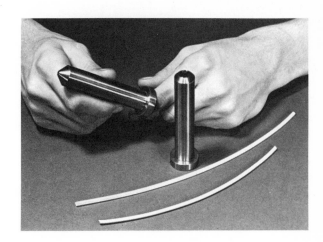

Fig. 6-25. These two extruder dies are for the production of round plastic rod. Dies of this type are easily constructed in the laboratory.

Fig. 6-26. The water-cooled take-off system for a laboratory extruder.

for particular shapes. In most cases, take-off systems are not supplied with the machine and must be constructed for the specific use of the extruder in the laboratory, Fig. 6-26.

The setup and operation of the extruder should be followed carefully for safe production. The extruder has two electrical components, Fig. 6-27, the heating devices and the power drive for the screw. The following procedure is suggested for the extrusion of low density polyethylene or other similar resins:

1. Turn on the power to the heating bands and adjust dials for specified temperature settings. (Follow resin manufacturer's specifications.)
2. Allow adequate warm-up time (until resin is in softened state). CAUTION:

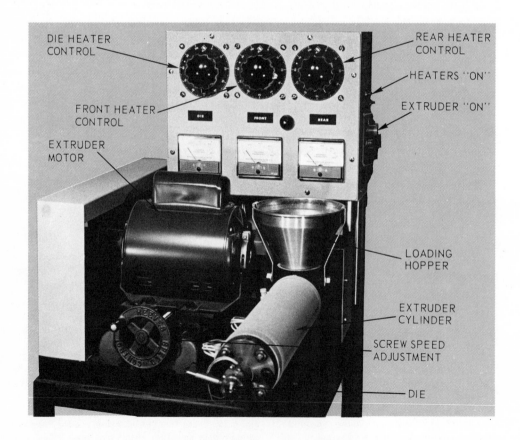

Fig. 6-27. A typical 3/4 in. laboratory extruder showing the operational features.

Extrusion

DO NOT TURN ON EXTRUDER POWER SWITCH WHEN RESIN IS SOLIDIFIED IN THE EXTRUSION CYLINDER AS THE SCREW MAY BREAK.

Fig. 6-28. Loading the extruder hopper with low density polyethylene granules for the production of 1/8 in. diameter welding rod.

3. When the correct temperatures for the heating units are obtained, fill the hopper with proper amount of resin for the particular run, Fig. 6-28.

Fig. 6-29. The extrudate is fed under the first hold down finger to begin the water cooling.

4. Fill the cooling tank with water to cover the hold down fingers.
5. Turn on the take-off rolls and adjust to an average speed.
6. Press extruder power switch and guide extrudate under hold down fingers as it begins to flow, Fig. 6-29.
7. When the extrudate approaches the end of the cooling tank, feed it between the take-off rolls, Fig. 6-30.

Fig. 6-30. The extrudate is fed into the take-off rolls as it comes out of the water. It is cool enough so it will not be distorted by the roll pressure.

8. Adjust the speed of the take-off rolls until there is a slight tension on the extrudate.
9. Cut the rod after it has been coiled, or at measured lengths. See Fig. 6-31.

Fig. 6-31. The 1/8 in. diameter rod is being cut to desired lengths.

10. When extrusion is completed, shut off extruder power, heaters, and take-off rolls.

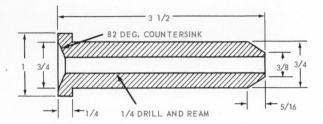

Fig. 6-32. A working drawing of a 1/4 in. diameter rod extrusion die.

Although an extrusion die may have accurate dimensions, the extrudate from the die must have a smaller cross section. This is due to the slight pulling of the take-off rolls. For example, a rod die of 1/4 in. diameter cannot produce a 1/4 in. diameter rod. The movement of the take-off rolls pulls the diameter down to less than 1/4 in. If tension is not placed on the extrudate, it will buckle and wiggle in the cooling tank.

The cross section of a typical extrusion die is shown in Fig. 6-32. A rod die of this type can be turned and bored on the lathe from mild steel. Other die shapes can be machined by using more than one piece of stock and fastening them together after finishing.

Sheet extrusion on laboratory size equipment is limited to narrow widths. Tape or ribbon may be extruded, Fig. 6-32A, and chill roll cooled before being wound up on a spool.

The blown flim process is best run in the laboratory with low density polyethylene, Fig. 6-32B. Bubble is pinched off at top by nip rollers which pull bubble upward, flatten and feed it to a wind-up spool. Film can then be used for packaging operations, Fig. 6-32C.

During trial runs and die testing, there is often considerable scrap plastic remaining. This material can be reground and returned to the extruder hopper or stored for later use. Care should be taken to keep the scrap and reground material clean.

EXTRUDER TAKE-OFF UNIT

Details on laboratory extruder take-off unit which can be constructed to meet the requirements of most extruders may be obtained from Figs. 6-33 and 6-34. The water cooling tank is made from a 4 ft. length of galvanized gutter spout with end caps soldered in place. The inside is coated with clear epoxy resin and the outside is painted. A drain plug is

Fig. 6-32A. A 4 in. wide extruder tape die and a sample of the resulting tape.

Fig. 6-32B. Photo shows small blown film bubble and film surface winder. (Filmaster Design Inc.)

Fig. 6-32C. Note use of blown film roll for a laboratory packaging operation.

Some profile shapes are too large or complex to be pulled through nip rollers. Caterpillar tread take-off units, Fig. 6-34A, have more surface area for pulling the extrudate. They also have the adjustability to handle larger cross sections.

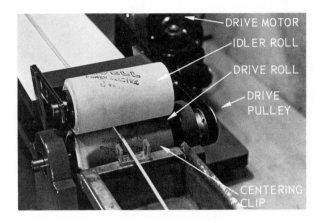

Fig. 6-34. Details of the drive and idler rolls, centering clip and motor.

provided for water removal. Hold down fingers of soft sheet aluminum adjust to the necessary depth requirements.

The take-off rolls are made from a washing machine wringer cut in half. The lower drive roll is mounted in ball bearings while the upper idler roll is hinged with pins, and provides enough weight to keep the plastic from slipping. Speed of a 1/6 hp motor, belted to the lower roll, is controlled by a wall mounted rheostat. An adjustable acrylic guide clip keeps the extrudate centered and the Formica topped outfeed table provides a good trimming surface. A range of 8 to 16 rpm for the take-off rolls will meet most extruding requirements.

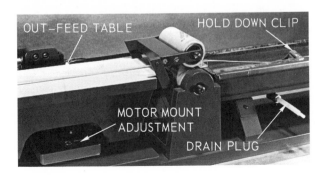

Fig. 6-33. Take-off system including tank drain, motor adjustment, and out-feed table.

Fig. 6-34A. This is an adjustable caterpillar tread take-off unit. (Killion Extruders, Inc.)

TEST YOUR KNOWLEDGE - CHAPTER 6

1. In a resin compounding plant, the extruder is used for:
   a._____.
   b._____.
   c._____.
2. The extrusion process is similar to _____.
3. The size of an extruder is measured by:
   a. The diameter of the screw.
   b. The length of the screw.
   c. The amount of resin extruded in one minute.
   d. None of the above.
4. Extrusion requires_____ and _____to cause the plastic to flow through the die.
5. List the three main classifications of extruded products.
   a._____.
   b._____.
   c._____.
6. Heat to soften the plastic granules in the extruder has two sources.
   a._____.
   b._____.
7. Formulating colored resin is done by _____ or _____

8. to the natural granules.
9. Four plastics widely used for wire coating are:
   a._____.
   b._____.
   c._____.
   d._____.
9. Extruded film material over_____inches thick is classified as sheet.
10. The device in an extruder used to remove dirt is called a _____.
11. What plastics are used in extrusion processes, thermoplastics or thermosetting resins?
12. The three main parts of an extruder take-off system for profile shapes are:
   a._____.
   b._____.
   c._____.
13. The major precaution to take when operating the laboratory extruder is to _____.
14. The part of the extruder that controls the shape of the extruded part is the _____.
15. The most commonly coextruded product is _____.

SUGGESTED ACTIVITIES

1. Using 1/8 or 3/16 in. rod die, extrude polypropylene or polyethylene welding rod to be used in hot gas welding.
2. Make a display board illustrating as many different cross section profiles of extruded products as you can find. Identify the resin from which each product was made by running a burning test.
3. Make a drawing of the extruder in your laboratory showing the extruder screw, cylinder, and die in cross section. Iden-

tify all parts.
4. Design and machine a die for the extruder in your laboratory. Test the die on a short production run.
5. Make a survey of literature on extruders and write a report on the latest innovations in extruder design and processing techniques. Include a listing of the resins most used in extruding today. List typical products for which they are being used.

Chapter 7
BLOW MOLDING

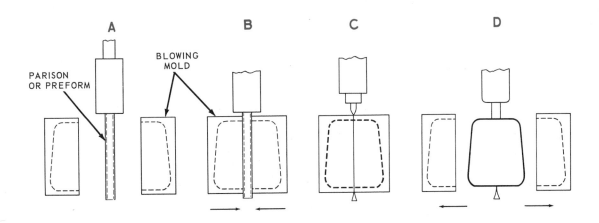

Fig. 7-1. Drawing of blow molding sequence: A—Molten hollow tube (parison) placed between the two halves of mold. B—Mold closes around parison. C—Parison, still molten, is pinched off and inflated by air blast which forces walls against inside contours of cold mold. D—When the piece has cooled enough to become solid, the mold is opened and the product ejected.

Blow molding is a process used to produce thin-wall hollow thermoplastic parts. A cylinder or tube of plastic, called a PARISON, is placed between the jaws of a mold. The mold is closed to pinch off the ends of the heated plastic and compressed air pressure

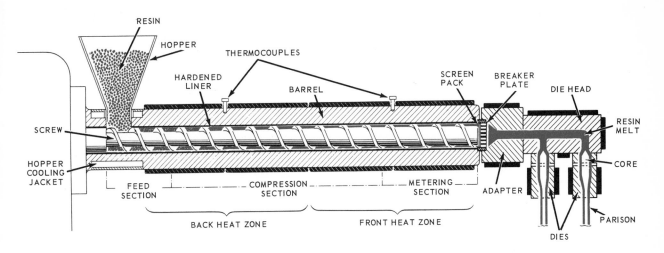

Fig. 7-2. Cross section of extrusion and die units of twin-head blow molder. (U.S. Industrial Chemicals)

is used to force the material against the mold faces. When cool, the plastic becomes rigid. See Figs. 7-1, and 7-2.

The three main phases of the commercial process are:
1. Softening the resin by the use of heat.
2. Forming the parison.
3. Blowing the parison in the mold.

The first of these phases involves the use of an extruder to heat the plastic to a molten state and compress the material at the die head. This part of the process is the same as typical extrusion. Second, the die, or multiple dies, form the diameter and wall thickness of the extruded parison which is ready to be clamped between the mold halves, Fig. 7-2. The third phase involves the closing of the mold halves by hydraulic pressure and pinching the parison. Air is used to expand the resin to conform to the mold cavity.

Fig. 7-4. This typewriter case is blow molded in one piece with double walls and a living hinge.
(Blowmolding Machinery Div., Hoover Universal Inc.)

Polyethylene resin is used extensively in blow molding. It is ideal for making a variety of products from soft, flexible squeeze bottles, to rigid containers.

Fig. 7-3. Blow molding was originally developed to produce simple round containers.

Fig. 7-5. Blow molded products may take intricate shapes such as this hollow polyethylene ski sled.   (Chemplex Co.)

PROCESS IS USED EXTENSIVELY

In recent years blow molding has become one of the major processing methods of the plastics industry. The process was originally developed to produce simple round containers as shown in Fig. 7-3. With the rapid advancement of blow molding technology and machine design, the process is now to the stage of high-speed production with unique product design possibilities. See Figs. 7-4 and 7-5. Almost any hollow object can be successfully blow molded, from chair seats and backs to automobile arm rests and sun visors.

Fig. 7-6. Disposable blow molded bottles are used to contain various liquid and powder materials.
(Hayssen Manufacturing Co.)

Fig. 7-7. Four head blow molding machine can make milk containers in less than a 10 second cycle.
(Blowmolding Machinery Div., Hoover Universal Inc.)

Most thermoplastics can be blow molded. Ionomer, polyvinyl chloride, polycarbonate, and acetal resins are but a few that are used in considerable quantity. Perhaps the greatest single use of blow molding is in the production of disposable containers, such as the bottles shown in Fig. 7-6. They are lightweight, nonbreakable, and easily disposed of by incineration.

The size of a blow molding machine is determined by the extruder screw diameter, the number of die heads, and the size molds the machine will take, Fig. 7-7. Blow molders are usually designed with the die heads lined up in a straight row with the molds on a fixed table, or a single die head on a multiple station mold turntable, Fig. 7-8.

Of major importance in understanding the blow molding operation are the details of a typical mold as shown in Fig. 7-9. The parting line on the product appears where the mold halves meet. Those sections of the mold that squeeze the parison and weld it together prior to blowing are known as pinch-offs. The sections of the containers that have been pinched off, Fig. 7-10, are removed later during the trimming operation. The section pinched off at the bottom

of a container is known as the tail. Machined aluminum is now the prime material for molds. Previously, extensive use was made of beryllium copper for the construction of the molds used in the blow molding process.

Fig. 7-8. Blow molding machine with rotating mold table. As blow molding machine table revolves, each mold performs a different operation. One receives the parison, moves to the blowing station, cooling station, and ejection station.

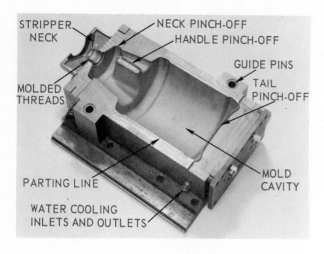

Fig. 7-9. One-half of mold for container with handle and threaded top.
(Blowmolding Machinery Div., Hoover Universal Inc.)

platens, Fig. 7-11. The die, Fig. 7-12, is adjusted to extrude a parison of the estimated wall thickness, and the hopper is loaded with plastic granules either by hand or through an automatic vacuum system. The heaters are turned on for a warm-up period which softens the resin in the extruder cylinder and the die heads. Air pressure is adjusted for blowing the container and operating the strippers which eject the product. The water cooling system is turned on to maintain correct mold temperature and the hydraulic system is activated. This opens and closes the molds. The extruder is then turned on, set on automatic cycle,

Fig. 7-10. These fabric softener bottles, just ejected from the machine, still have the pinch-off sections attached.
(Blowmolding Machinery Div., Hoover Universal Inc.)

Fig. 7-11. The multicavity blow mold is mounted to the movable platens.
(Blowmolding Machinery Div., Hoover Universal Inc.)

THE MOLDING OPERATION

The sequence of the blow molding operation begins with the securing of the required molds onto the movable hydraulic

and a parison, Fig. 7-13, is extruded between the open halves of a mold as shown in Fig. 7-14. After the blowing cycle, the mold halves open and the container is ejected, Fig. 7-15. In some operations, the container is conveyed to an automatic trimmer, Fig.

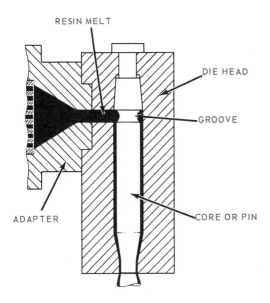

Fig. 7-12. Schematic drawing of side fed die with grooved core. The core is adjustable to give the required wall thickness to the parison. (U.S. Industrial Chemicals)

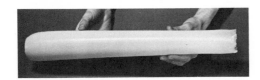

Fig. 7-13. A parison removed from the machine before molding shows a slight sag due to the weight of the resin.

Fig. 7-14. The parison approaches the bottom of its cycle between the open molds. This is almost clear, due to molten state of the plastic. A container is being blown in the closed mold at the right.

Fig. 7-15. Mold has opened and container is ejected by the stripper. Pinched off tail is at bottom of container.

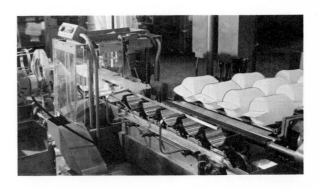

Fig. 7-15A. Plastic bleach bottles are conveyed to automatic trimmer where neck and tail pinch-offs are removed. (Blowmolding Machinery Div., Hoover Universal Inc.)

7-15A, while in others, the trimming operation is separate as in Fig. 7-16. As with most other thermoplastic processes, the scrap from trimming and defective containers can be reground, Fig. 7-17, and returned to the hopper to be used again. This sequence of the blow molding procedure, with variations, is characteristic of the procedure used in forming most blow molded products, Fig. 7-18.

INJECTION BLOW MOLDING

In the injection blow molding process, the parison is injection molded instead of ex-

Fig. 7-16. The handle section broken off by the trimmer. Reamer has just faced off the neck so the cap will seal tightly.

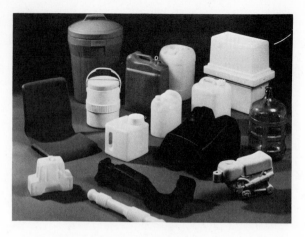

Fig. 7-18. Note industrial containers and parts produced by blow molding.
(Blowmolding Machinery Div., Hoover Universal Inc.)

Fig. 7-17. This thermoplastic regrinder shows the cutting knives which reduce the plastics to flakes for further processing.

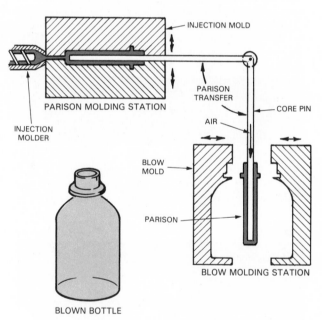

Fig. 7-18A. Study this diagram of the injection blow molding process.

truded. The parison is molded around a core pin then transferred and clamped into a blow mold, Fig. 7-18A. At this station, air is forced through the core pin to expand the hot parison into the mold cavity.

The finished product has no flash or waste since the parison was not pinched off at the ends, Fig. 7-18B. Accurate neck shapes are achieved due to the injection molding influence on that section of the bottle. Irregular shaped

containers can be molded by varying the wall thickness of the parison.

Injection blow molding equipment, Fig. 7-18C, is more expensive than traditional machinery because two molds and molding stations are required. Also, it is not as fast

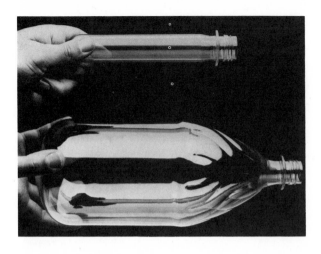

Fig. 7-18B. An injection molded parison and the resulting blown bottle, note wall uniformity and detail of neck. (Cincinnati Milacron Co.)

Fig. 7-19. Laboratory blow molding unit has stationary and movable mold halves, guide bars, and clamping device.

Fig. 7-18C. These carbonated beverage containers are being examined on an injection blow molding line. (Cincinnati Milacron Co.)

a process and has more limitations on part complexity and size.

BLOW MOLDING IN THE LABORATORY

A laboratory size blow molding unit, Fig. 7-19, can be constructed by the instructor or students in an advanced course. Many design possibilities are adaptable to this type of unit, and the type to make depends on the shape of the required product. In that it should follow the principles of industrial blow molding, the unit should involve a split mold and at least one movable mold half. Fig. 7-20 indicates construction details on the mold shown in Fig. 7-19.

A number of materials are suitable for mold construction. Wood and cast or machined aluminum are easy to work and give good detail. The greatest problem in mold construction is the machining and polishing of the mold cavity, Fig. 7-21, as an internal shape is involved. If a cast aluminum mold is to be made, the normal procedures of green sand casting may be used with a wood pattern of the container. After casting, the mold faces should be machined flat and the cavity carefully polished.

In constructing the mold for the bottle, Fig. 7-20, the procedure is:

1. Prepare two blocks of hard maple of desired size for the mold halves.

2. Clamp the blocks together and bore dowel holes through one-half and at least one inch into the second half. The holes will be for the guide bars which keep the mold in alignment.

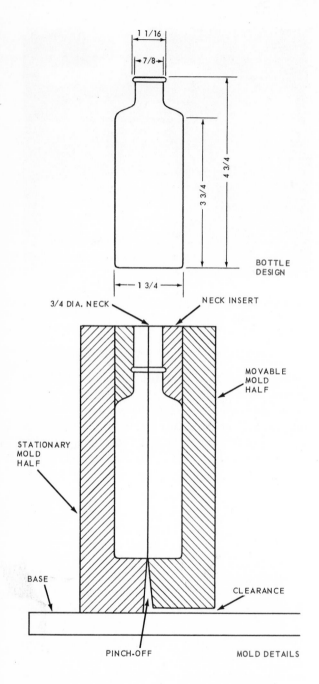

1 1/16
7/8
4 3/4
3 3/4
BOTTLE DESIGN
1 3/4
3/4 DIA. NECK
NECK INSERT
MOVABLE MOLD HALF
STATIONARY MOLD HALF
BASE
CLEARANCE
PINCH-OFF
MOLD DETAILS

Fig. 7-20. Above. Design details for a blow molded bottle. Below. A cross section of the blowing mold.

3. With the mold halves in a vertical position, clamped together, bore the 1 3/4 in. diameter cavity body on the mold parting line to correct depth.

4. Make a split turning for the cavity neck insert, separate, and glue to the respective mold halves.

Fig. 7-21. The mold in open position shows the neck, cavity, and tail pinch-off.

5. Carefully sand the cavity and machine the pinch-off on each mold half to 1/16 in. separation to provide for the tail weld.

6. Apply a coat of clear epoxy to the cavity and pinch-off surfaces, sand lightly with wet-or-dry abrasive paper.

Fig. 7-22. A copper wire fastened to the parison with a hook at the upper end, makes it easy to hang in oven for heating.

Fig. 7-23. Placing the molten parison against the neck of the stationary mold half before closing the mold.

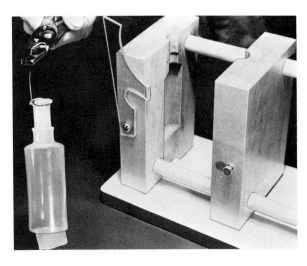

Fig. 7-25. Removal of the blown bottle from the mold. Pliers should be used as the neck of the bottle may still be hot.

7. Cover the exterior of the mold with two coats of clear lacquer.

8. Glue the guide bars into the stationary mold half and secure it to the base with screws.

9. Wax the guide bars and trim enough off the movable mold half so it will clear the base.

10. Design and attach mold clamping device.

BLOWING PROCEDURE

A number of resins may be used for laboratory blow molding, however, it has been found that high density polyethylene, either natural or colored, provides good results because of its rigidity and ease of processing. The parison is cut from 3/4 in.

Fig. 7-24. The air nozzle seals against the softened plastic as the parison is inflated.

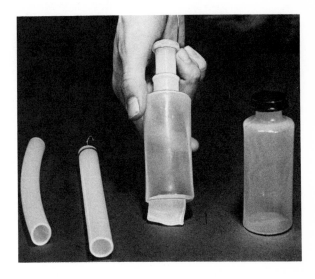

Fig. 7-26. A finished molded bottle on the right along with an untrimmed bottle and sections of tubing made ready for heating.

diameter tubing having a 1/16 in. wall thickness. Thicker tubing should be used for larger products. A copper wire is secured to one end of the parison and it is hung in the heating oven set at 325 deg. F., Fig. 7-22. Heating should take from 3 to 5 min. at which time the parison is quickly removed from the oven and placed between the mold halves of the blow molding unit as shown in Fig. 7-23. The movable mold half is closed on the parison, locked tightly in position by the clamps, and air is blown in through the neck, Fig. 7-24. Air pressure between 30 and 40 psi is satisfactory as the nozzle of the air gun seals at the neck of the molten parison. Pressure should be maintained for a few seconds, the mold opened, and the product removed as in Fig. 7-25. The mold is ready for another molding cycle immediately. Upon removal from the mold, the tail can be broken off the bottle and the neck trimmed with a sharp knife. The finished blow molded bottle is shown in Fig. 7-26. An injection molded low density polyethylene snap cap may also be made for a blow molded container. See page 95 for molded cap designs.

MACHINE OPERATION

Laboratory blow molding machines consist of a basic extruder with a tubular die. The blow mold is located directly beneath the die. It is generally made of cast or machined aluminum. Metal filled epoxy may also be used.

The process begins by heating the extruder and loading it with plastic granules. The parison is extruded down between the two mold halves, Fig. 7-27A. The mold is manually clamped shut, B, pinching and sealing the tail of the parison. When the air switch is turned, air expands the bottle within the mold, C. A short cooling period takes place before the mold is opened, D. The bottle is then hand stripped from the mold. Neck and tail are trimmed with a knife.

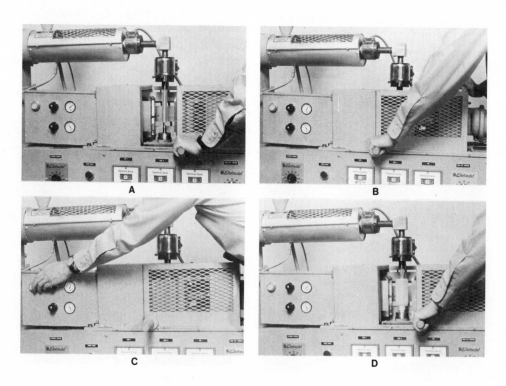

Fig. 7-27. A—The parison is extruded. B—The mold is clamped around the parison. C—Air is blown into parison, expanding it into the mold cavity. D—The mold is opened, revealing the finished product.
(Kissam Manufacturing, Inc.)

Blow Molding

1. A polymer used in blow molding is _____.
2. The tube of hot plastic extruded between the molds of a blow molder is called a _____.
3. The sections of a blown product where parts have been squeezed by the mold are known as _____.
4. Three operations in the blow molding process are:
   a. _____.
   b. _____.
   c. _____.
5. The size of a blow molding machine is determined by:
   a. _____.
   b. _____.
   c. _____.
6. The section that is removed from the bottom of a blown container is known as the _____.
7. What is the purpose of the stripper on a blow molding machine?
8. After most blow molding operations, the product must be:
   a. Polished.
   b. Cleaned.
   c. Trimmed.
   d. Tempered.
9. _____ is used to maintain cool molds during the blow molding process.
10. What is done with defective parts and scrap from blow molding?
11. What kind of objects may be sucessfully blow molded?
12. The mold halves produce a mark on the blow molded product called the _____.
13. The metal most used for molds is:
    a. Aluminum.
    b. Steel.
    c. Brass.
    d. Bronze.
    e. None of the above.
14. In commercial molding, a _____ is used to load the plastic granules into the hopper.
15. Blow molding was originally developed to produce _____.
16. Injection blow molded products are made with no _____ or _____.

SUGGESTED ACTIVITIES

1. Make a display for your laboratory of containers and other blow molded products. Indicate the type of resin used for each product and describe why it was selected.
2. Design and construct a molding unit for blow molding in your laboratory. Test the mold for production. If a molding unit is available, blow mold a number of products using suitable resins.
3. Make a collection of hollow containers such as squeeze bottles and detergent bottles. Examine each container to determine if it was blow molded by checking these factors:
   a. Look for a weld line at the bottom of the container where the tail may have been pinched off.
   b. Check the neck area of the container where trimming may have been necessary.
   c. See if the neck has been machined to seat the cap.
   d. Check for a parting line around the product where the mold halves came together.
4. Design and construct a fixture for cutting tubing to the correct length for blow molding in your laboratory.
5. Investigate recent literature and make a report on the latest innovations in commercial blow molding. Include pictures or sketches of new processes.
6. Visit a plastics company to observe blow molding on a production scale. Relate your observations to this chapter.

This is a 150 ton hydraulic press with molds attached for compression molding.   (Dake Corp.)

Chapter 8
COMPRESSION AND TRANSFER MOLDING

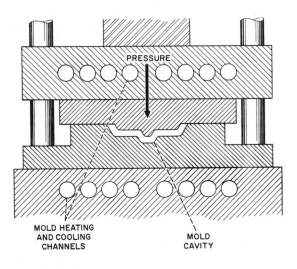

Fig. 8-1. Schematic cross section of a typical single cavity compression mold. (U.S. Industrial Chemicals)

COMPRESSION MOLDING

In compression molding, a measured quantity of thermosetting resin is placed in a mold. The mold is heated, pressure is applied, and the molten resin completely fills the mold cavity. See Fig. 8-1. The resin undergoes a chemical reaction while hardening to a permanent form.

Molding materials commonly used are: phenolic, alkyd resin, diallyl phthalate, melamine, and urea, Figs. 8-2 and 8-2A. These resins are available in the form of fine powders, granules, flakes and rope. They are also available as PREFORMS (in a preform the specified amount of resin is compressed and

Fig. 8-2. These are compression and transfer molded phenolic appliance parts.
(Hooker Chemicals & Plastics Corp.)

Fig. 8-2A. Compression molded melamine dinnerware is hard, heat resistant and durable. (Heller Designs, Inc.)

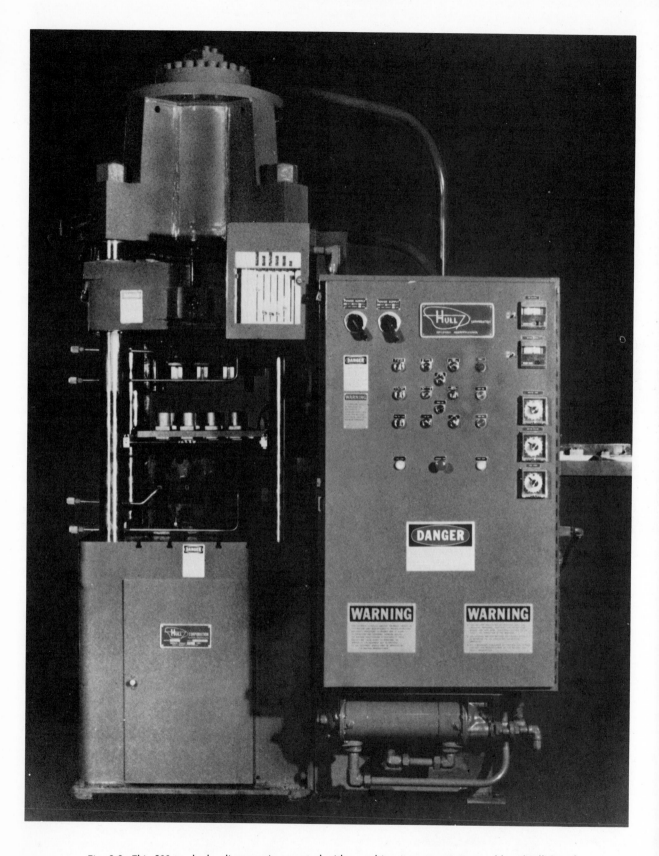

Fig. 8-3. This 200 ton hydraulic press is mounted with a multicavity compression mold.   (Hull Corp.)

sticks together like an aspirin tablet). A preform contains just enough resin to fill the mold and to allow for a slight flash (extra plastic attached to molding along parting line which must be removed before the part is completed). Plastic pastes or rope preweighed into slugs, are used the same way.

In compression molding a charge of resin is placed in an open two-part heated mold. The mold is then closed. Pressures from 1,000 to 10,000 psi are applied by a hydraulic press, Fig. 8-3. Mold temperatures which range from 280 to 400 deg. F., cause the resin to melt and flow through the openings of the mold cavity. When polymerization is complete, which takes from 3 to 20 min., the mold is opened and the product is removed. Mold temperatures can be maintained and the product removed hot as the thermosetting material is in a solid state. The next charge can be placed in the mold immediately.

Fig. 8-5. The compression molding machine operator prepares and weighs slugs of diallyl phthalate preform compound to be placed in the mold by hand.

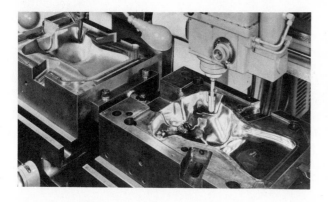

Fig. 8-4. One-half of a compression mold being machined on an automatic tracer milling machine. The mold model is at the right.   (Cincinnati Milacron Co.)

charge, loads the molds, removes the molded product, Fig. 8-6, and often trims the flash during the cure period. Compression molding done on automatic presses, Fig. 8-7, requires a cycle in which the molding com-

Fig. 8-6. After a five minute curing period, the mold opens and the machine operator removes a molded aircraft circuit breaker made from diallyl phthalate.   (FMC Corp.)

The mold for compression molding can be a single or multiple cavity type. Molds are usually made of tool steel and are highly polished to produce the desired product finish, Fig. 8-4. The mold halves are bolted to the upper and lower platens of the hydraulic press for automatic opening and closing during the molding cycle.

If the molding process is manually operated, Fig. 8-5, the operator prepares the

pound is fed directly into each mold cavity, compressed, cured, and the product ejected from the press. If inserts are required in the product, Fig. 8-8, they are usually placed in the open mold by the machine operator.

Fig. 8-7. Automatic compression molding machines producing wall outlet covers of urea resin. (Dake Corp.)

TRANSFER MOLDING

Transfer molding involves the same principles and materials as compression molding. The main difference between the two processes is in how the material is charged into the cavity. See Fig. 8-9. In pot type transfer molding the resin is not fed directly into the mold cavity, but into a separate chamber where it is heated under the pressure of a plunger until molten. Higher pressures are then exerted on the plunger, from 6,000 to 12,000 psi, forcing the softened resin through the runners and gates into the mold cavities, Fig. 8-10. This part of the process is similar to injection molding of thermoplastics.

Fig. 8-8. This coil top and distributor cap made of phenolic, require metal inserts in the compression mold before loading. (Hooker Chemicals and Plastics Corp.)

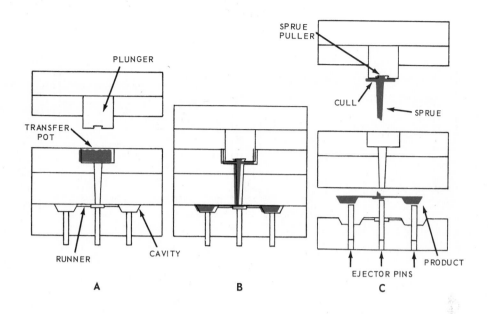

Fig. 8-9. Schematic drawing of the transfer molding cycle. A—Mold open and resin placed in transfer pot. B—Plunger forces molten resin through the sprue into the mold cavities. C—As the mold opens, the cull and sprue break loose and the part is removed by ejector pins.

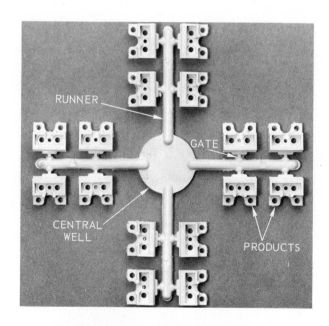

Fig. 8-10. The finished product ejected from a transfer mold ready for trimming.   (American Cyanamid Co.)

High-speed plunger molding uses an auxiliary transfer ram to force plastic through runner and into cavities, Fig. 8-11. Transfer pressure is more controllable since it is independent of clamping pressure. The auxiliary ram is usually at the top of the press. It works downward through the upper platen. High-speed plunger molding is generally faster than the pot type; however, pot type molds can be used on standard compression presses.

The molding compound is usually preheated to speed up the molding cycle. Only one shot can be made at a time and all the polymerized material, the cull, sprue, runners, and products, must be removed before the next cycle. All parts of the molding system, except the products being molded, become scrap since the thermosetting materials cannot be resoftened.

Transfer molding has the advantage over compression molding in that no flash is involved, which requires less finishing. It also makes it possible to mold many products at the same time through the use of a runner system. It is especially well suited to the molding of small intricate parts that would be difficult to compression mold, as shown in Fig. 8-12.

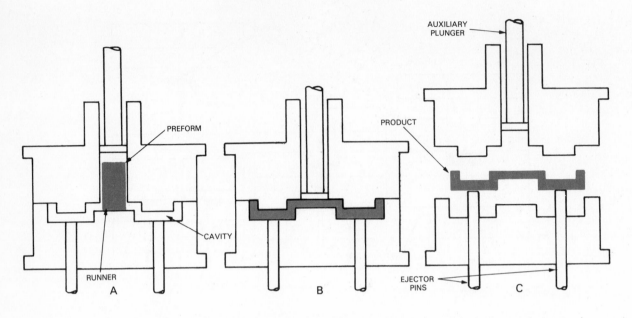

Fig. 8-11. Study diagram of high-speed plunger transfer molding. A—Preheating the preform. B—Molding. C—Ejecting parts.

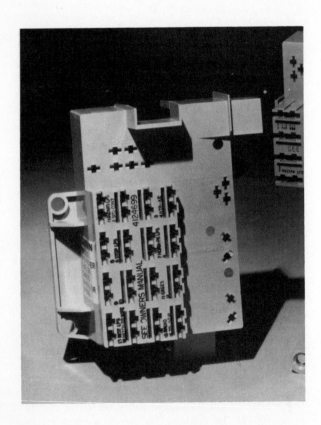

Fig. 8-12. These are electrical components transfer molded with diallyl phthalate around delicate conductive inserts.

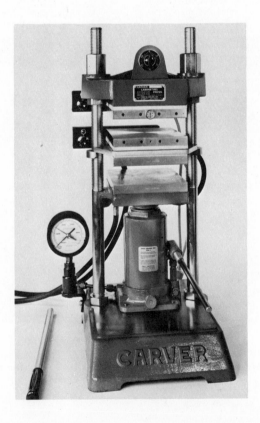

Fig. 8-13. This manually operated hydraulic press has electrically heated platens, is water cooled, and contains a temperature scale and controls.   (Fred S. Carver, Inc.)

COMPRESSION MOLDING IN LABORATORY

The plastics laboratory provides an excellent opportunity to carry out experimentation and production of products by the compression molding process.

A heated platen, hydraulic press, as shown in Fig. 8-13, is quite a versatile type of machine for compression molding. If more than one press is available, the molds can be mounted directly to the platens. However, this often limits the press for use in other processes. The press should have a heating capacity of at least 400 deg. F., and deliver 4,000 psi pressure. In metric values, this equals 205 deg. C and 27560 kPa.

Fig. 8-14. Parts of compression mold for a cabinet knob.
(Dake Corp.)

MOLD CONSTRUCTION

Many different mold designs can be developed for compression molding. A mild steel is appropriate for construction. Mild steel is relatively easy to machine, polishes to a good finish, and conducts and holds heat well. A simple flash mold for a molded cabinet knob is shown in Fig. 8-14. The general procedure for developing this and other molds is:

1. Select a bar of mild steel of correct diameter and cut off two sections for the mold halves.
2. Chuck the stock in the lathe, turn to the correct diameter, and face off each end.

3. Machine the cavity in the bottom half of the mold. Polish. Bore hole for the insert which will hold a machine screw. This screw provides the threaded hole in the finished knob.

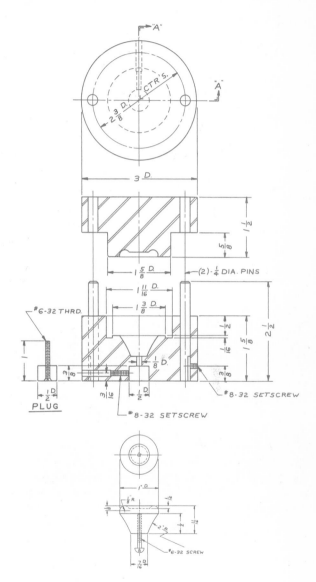

Fig. 8-15. Detailed drawing of the compression mold for a cabinet knob.

4. Drill and tap the hole for the setscrew to hold the plug in place.
5. Turn the cavity in the top half of the mold. Polish smooth.
6. Line up the mold halves, clamp them together, and drill holes for the guide pins.

7. Insert the guide pins so they are tight in the top half but slide smoothly into the lower half.

8. Stamp an index mark on each mold half. The mold can then be assembled with the mold halves lining up the same way for each cycle.

Fig. 8-17. Mold release has been sprayed on the mold and the top half is lined up with the guide pins as mold is closed.

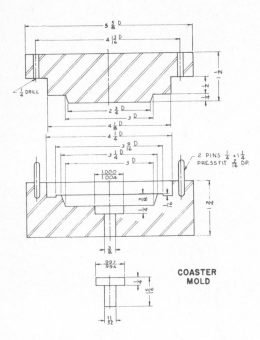

Fig. 8-16. Working drawing of a compression mold for a plastic coaster. (Dake Corp.)

Fig. 8-17, should be placed in the press during this warm-up period, making contact with both platens, so it will be at the correct temperature at the time of molding. For compression molding a coaster, the exact amount of molding compound is weighed, Fig. 8-18,

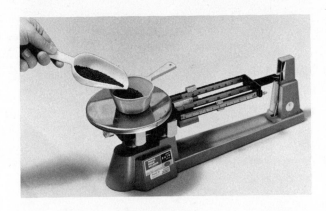

Fig. 8-18. Weighing the charge on a triple beam balance. The weight of the cup, 14 grams, plus the correct charge of 21 grams has been set on the balance at 35 grams. Phenolic resin is added until the pointer is on zero.

Fig. 8-15 provides a working drawing of the mold for the compression molded knob. A similar drawing of the mold for a plastic coaster is shown in Fig. 8-16. Other possible compression molded products would include the base for a desk pen set, small parts containers, tool handles, threaded bottle caps, and electrical insulator bases.

MOLDING PROCEDURE

Compression molding requires a number of steps which should be followed in producing a quality product. The heating platens of the molding press should be turned on at least 15 min. prior to use and set at the appropriate temperature for the molding compound to be used. The assembled mold,

and made ready for loading into the mold. In this case, wood flour filled phenolic is being used. When the required mold temperature is attained, the mold is removed from the press and opened to receive the charge, Fig. 8-19. Heat resistant gloves should be used when handling the hot mold. The mold should be closed and returned to the press as quickly as possible to retain the mold temperature. Pressure is applied to the mold with the hydraulic pump handle and main-

Fig. 8-19. The phenolic charge is poured directly into the lower half of the hot mold.

Fig. 8-21. Tapping the ejector pin with a dowel rod to remove the molded coaster. A wood dowel rod is used so as not to damage the mold.

tained at 2,000 psi for 8 min. After curing, the mold is removed from the press, Fig. 8-20, and separated. The molded coaster is removed from the mold while it is still hot, Fig. 8-21, and the mold is prepared for the next cycle. When the coaster has cooled, the flash is carefully filed off and the product completed, Fig. 8-22.

be removed. This involves reheating the mold for each product and the lengthy cycle period usually makes it prohibitive for production purposes. However, a special acrylic resin produced as tiny spheres is made for compression molding. Its primary use is for embedding scientific specimens and items for display. The process is similar to the compression molding of thermosets and can be used in the laboratory for embedding coins, scientific samples and other similar items.

An all-purpose mold can be constructed which will serve to make moldings from 1/8 to 1 1/2 in. in thickness, see Fig. 8-23.

Fig. 8-20. With pressure released, the hot mold is removed from the press after the curing period.

Fig. 8-22. Filing the flash from the edge of the coaster. Completed coasters are shown at the right.

Little compression molding is done using thermoplastic materials, mainly because the mold must be cooled before the product can

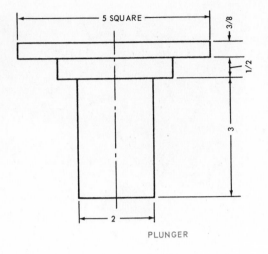

PLUNGER

Fig. 8-25. The embedded specimen is gently covered with acrylic resin. The plunger half of the mold is assembled with the base.

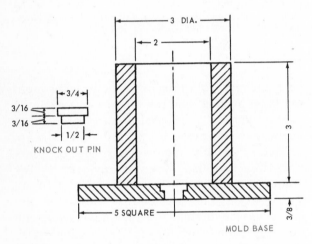

KNOCK OUT PIN

MOLD BASE

Fig. 8-23. Detail drawing of compression mold for acrylic embedment.

into the mold. Using tweezers, the specimen is gently placed in the center of the resin and the remainder of the acrylic compound is poured into the mold, as shown in Fig. 8-25. The specimen used here is a valuable coin. The plunger half of the mold is assembled with the base and inserted in the press at 310 deg. F. Pressure is slowly brought up to 500 psi, Fig. 8-26, and held for 10 min. During this period the acrylic resin melts and, being of such a fine bead,

As in regular compression molding, the specified amount of acrylic resin is weighed, Fig. 8-24, and half of the charge is poured

Fig. 8-24. A charge of acrylic molding compound being weighed to determine the exact quantity required.

Fig. 8-26. Pressure being applied to the mold in the compression press.

surrounds the specimen without distortion and without the entrapment of air bubbles. The mold is then cooled by turning on the water cooling system which passes through the press platens. The mold is removed from the press, the product ejected, and buffed to a high gloss, Fig. 8-27.

The compression molding process provides an opportunity for widespread study of industrial processing, resin characteristics and properties, product and mold design, and laboratory experimentation and production.

Fig. 8-27. The product, crystal clear, permanently embedding the specimen without distortion.

TEST YOUR KNOWLEDGE - CHAPTER 8

1. Compression and transfer molding are used primarily with _____ resins.
2. Three advantages of transfer molding over compression molding are:
   a._____.
   b._____.
   c._____.
3. Resins for compression and transfer molding are produced in the form of:
   a._____.
   b._____.
   c._____.
   d._____.
4. Molding material of the correct amount and pressure formed as an aspirin tablet is a_____.
5. Five common polymers used in compression and transfer molding processes are:
   a._____.
   b._____.
   c._____.
   d._____.
   e._____.
6. What two factors cause the resin to become molten in these processes?
   a._____.
   b._____.
7. Plastic rope is preweighed into _____ before being placed in the mold.
8. What is done with the scrap from these molding processes? Why?
9. In transfer molding, the resin is placed in a _____ to be softened before molding.
10. Compression molding pressures range from _____ to _____ psi and are applied by_____ pressure.
11. Compression and transfer molds are normally made of_____.

SUGGESTED ACTIVITIES

1. Design a product and make a steel mold for compression molding the product. Test the mold using a variety of thermosetting molding compounds.
2. Select an article to be embedded and compression mold it in acrylic molding compound. Follow the manufacturer's specifications for molding with the acrylic beads.
3. Write a detailed report on the various fillers used in thermosetting resins for compression and transfer molding. Indicate why each filler is used and what properties it adds to the resin.
4. Make a display, using pictures and products, that shows the variety of uses of compression and transfer molded materials. Indicate what properties make

the particular molding compound ideal for each product application.

5. Collect samples of as many thermosetting molding compounds as you can find. Design a display board around the samples which may be heat sealed in plastic.

6. Visit a local plastic molding company and observe the procedures they use in compression and transfer molding. Write an outline report covering their method of molding for the products they manufacture.

7. Design and construct a mold for making preforms for a compression mold in your laboratory. Be sure the mold is designed to produce a preform that contains the exact amount of molding compound for the product involved.

8. Select a thermosetting resin used in compression or transfer molding and make an in-depth study of the polymer from the raw material stage to final products.

9. Using a mold from your laboratory, make a limited production run of the compression molded product. Make a production schedule which includes cycle times and volume of material used.

Calendering was used to produce many of the decorative coverings in this room. The upholstery for the couch, pillows and chair as well as woodgrain paneling were made by the calendering process.

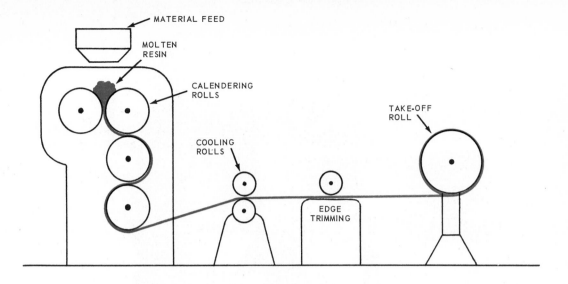

Fig. 9-1. Schematic illustrating the main stages in the calendering of polyvinyl chloride film.

Chapter 9
CALENDERING

Calendering is the process of squeezing a softened thermoplastic material between two or more rolls to form a continuous film. The process, originally adapted from the rubber industry, is a major method of producing plastic film and sheet.

The material generally used in the calendering process is flexible polyvinyl chloride. However, some film is calendered from ABS, cellulosics, polyethylene, and polystyrene.

Fig. 9-1 shows the stages in the calendering process. The thermoplastic resin may be mixed with lubricants, stabilizers, plasticizers, and colorants. During mixing, the mass is heated and becomes rubbery, like hot, soft clay. It is then fed into heated calendering rolls, and squeezed to the desired

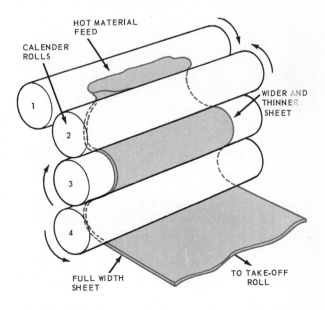

Fig. 9-2. Widening and thinning of calendered sheet as it passes through the rolls.

139

Fig. 9-3. This huge calender has four rolls 69 in. (175 cm) wide. It is called a four-roll inverted L because of the location of the rolls.   (Steward Bolling and Co.)

thickness as it passes through the rolls. Passing from the calendering rolls, the vinyl film goes through cooling rolls, is cut to width by the edge trimmer, and is wound on a take-off roll. Fig. 9-2 shows the resin flow as it passes through the calendering rolls. Thickness of the calendered sheet is closely controlled by the space between the final rolls.

CALENDERING EQUIPMENT

The equipment used in calendering consists of four main units. A mixer or mill is used to blend the ingredients into the proper compound for processing. The major unit is

Fig. 9-4. A small two-roll calender having a maximum width of 16 in.

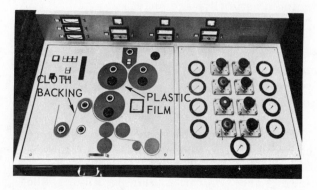

Fig. 9-5. The control panel for a coating calender shows the fabric backing meeting the plastic film and bonding between steel rolls.   (C. Tennant, Sons and Co.)

the calendering machine, Fig. 9-3, which compresses and rolls the plastic into a flat sheet or film. Finishing equipment includes cooling rolls and an edge trimmer. Finally, after finishing, the sheet is wound onto a take-off roll.

The calender itself serves to slowly reduce the plastic thickness as each set of rolls is spaced closer together. The rolls are heated to maintain forming temperature and are powered to draw the plastic through. Patterned or textured sheet is produced by embossed rolls. A small calendering machine, suitable for laboratory use, is shown in Fig. 9-4.

THE COATING PROCESS

Calenders are also used to coat sheet materials such as paper and fabric. The process is similar to regular calendering except fabric is fed into the calendering rolls as the plastic sheet is being formed, Fig. 9-5. Hot and soft plastic is forced against the fabric. The plastic and fabric are tightly bonded together, emerging as a single sheet through the final rolls. See Fig. 9-6.

Calendering is used to produce products such as furniture and automotive upholstery, Fig. 9-7, clothing, footwear, luggage and handbags. These products are made from flexible polyvinyl sheet as a coating over fabric. Flexible PVC film and sheeting, without backing or support are used for swimming pool liners, shower curtains, rainwear, simulated leather, auto crash pads and hard tops, Fig. 9-8. Rigid PVC is used for blister packaging, credit cards, sound tapes, lighting fixtures and woodgrain laminate covers.

Fig. 9-7. Calendered auto upholstery, a familiar use of flexible polyvinyl chloride, must withstand long exposure to ultraviolet light and temperature extremes.
(The B.F. Goodrich Co.)

Fig. 9-6 A large calender for coating paper and fabrics. The control panel is shown at the right.

Fig. 9-8. Calendered leatherette texture sheet is used as a replacement for real leather in crafts applications.
(General Electric Co.)

TEST YOUR KNOWLEDGE – CHAPTER 9

1. The four main units in calendering equipment are:
   a._____ .
   b._____ .
   c._____ .
   d._____ .
2. Calendering is used to produce plastic _____ and _____ .
3. Common plastic resins used in the calendering process are:
   a._____ .
   b._____ .
   c._____ .
   d._____ .
   e._____ .
4. Calendering rolls are_____ to maintain an even drawing down of the plastic thickness.
5. Patterned material is produced by using _____rolls.
6. The calendering process was originally adapted from the_____ industry.
7. Materials such as_____ and _____ are coated by the calendering process.
8. Five common products made by the calendering process are:
   a._____ .
   b._____ .
   c._____ .
   d._____ .
   e._____ .
9. The consistency of the hot plastic when calendering begins is like_____ .

SUGGESTED ACTIVITIES

1. From observation at a local hardware store, make a written list of the plastic products you see that are made by the calendering process.
2. Write to manufacturers of calendering machines and ask for literature on their equipment. Make a bulletin board display from the material you receive which illustrates the calendering process.
3. Make a collection of film and sheet plastics, plain and coated. Try to determine if each was manufactured by the calendering process. Perform a burning identification test on a piece of each sample.
4. Develop and produce a model of a four-roll calendering machine. Use a sheet of flexible plastic to run through the working model which will demonstrate the process.
5. Prepare a chart which shows the ingredients of a typical polyvinyl chloride compound for calendering. Explain the purpose of each of the ingredients used.
6. Write a short report on the plastic resins used in calendering at the present time. Include an estimate of the amount of each resin used for this process yearly.

Chapter 10
POWDER MOLDING
AND COATING

Powder Molding and Powder Coating are manufacturing processes which make use of dry plastic powders.

In POWDER MOLDING, the powder is used in a manner similar to blow molding or injection molding with plastic granules. It may be used to form either a solid or a hollow product.

In POWDER COATING, powdered resin is fused around all or part of a given substrate (a shaped or formed base) which may be metal, glass, ceramic, or other material.

In powder molding and coating, a number of plastics are used, the leaders being polyvinyl chloride, polyethylene, nylon and epoxy. Polyvinyl chloride has outstanding toughness, durability, and electrical insulating properties. Polyethylene powders offer excellent chemical resistance and low water absorption along with flexibility. Powdered nylon, although more expensive than most other plastics, provides ease of processing, hardness, toughness, and good abrasion resistance. Epoxy powder, the main thermosetting material used, is primarily selected for electrical insulating applications where hardness and high heat resistance are required.

ROTATIONAL MOLDING

Plastic powders, also plastic liquids, (see section on plastisol molding) may be formed by using rotational molding. Fig. 10-A1. Rotational molding involves a hollow mold which contains a charge of powdered resin. The

Fig. 10-A1. This 400 gallon (1514 litre) rotationally molded crop spraying tank is made from polyethylene powder. (ICI Americas Inc.)

mold is rotated biaxially (on two axes) and is mounted on an offset arm which spins the mold in two directions at the same time. See Fig. 10-1.

The process requires that the machine be designed to meet three major functions: a mold loading station, a heating oven, and a cooling station, Fig. 10-2. The powder is evenly spread over the mold surface. As the resin melts, it forms a solid coating on the inside of the mold to provide the required product shape.

Equipment for rotational molding is usually less costly and less complex than comparable injection or blow molding machines, however, production cycles are slower. A typical rotational molding machine is shown in Fig. 10-3.

143

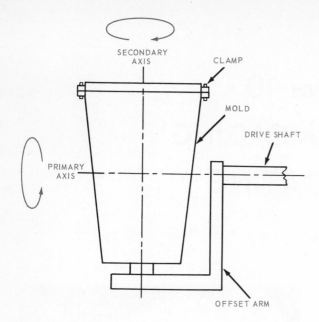

Fig. 10-1. A mounted mold on the offset arm rotates on two axes spreading the powdered resin evenly over the inner mold walls.

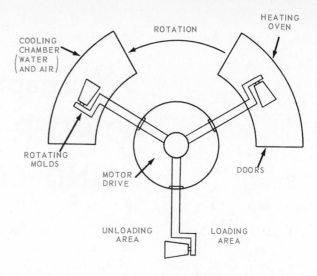

Fig. 10-2. Schematic top view of a three-mold rotational molder. The molds pass from the loading area on through the oven and cooling stations prior to product removal.

ROTATIONAL MOLDS

The molds for rotational molding are usually made of cast aluminum. This is easily machined, conducts heat and cold quickly, and is light weight. Molds too large to cast in aluminum are made of fabricated sheet steel. Since the molds require no water cooling holes and but little pressure is exerted on them, they can be made quite simply. The main requirements are that they be split molds for product removal, well polished for surface texture, and a well-sealed parting line, Fig. 10-4. Large molds are attached to the rotating arm and molded individually, Fig. 10-5. Smaller products can be molded in multiple cavity molds.

Fig. 10-3. A rotational molding machine with three spindles for attaching molds. This machine will hold three molds, each measuring 3 ft. square by 6 ft. high.   (McNeil Corp.)

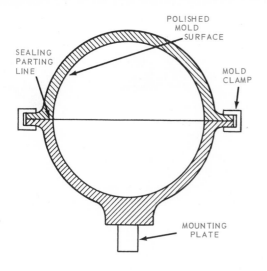

Fig. 10-4. Cross section of a cast aluminum mold for a hollow round ball. Parting surfaces must be flat to prevent flash.

An industrial container without a lid may be constructed by molding the complete container and cutting the necessary opening. Another method is to use an insulating cover on the mold where an opening is required. In the case of the container, a lid for the mold made of asbestos board or silicone foam will keep the powder from fusing at that area and an open end container will re-

sult. Mold release is used to prevent the product from sticking to the mold and making removal difficult.

MOLDING PROCESS

Six major steps are involved in the molding process:

1. The cavities of the mold are loaded with a measured amount of powder to produce the correct wall thickness in the product, Fig. 10-6.

2. The mold halves are clamped together tightly.

3. The charged molds are placed in the heating oven and rotated biaxially for the fusion period.

4. After all of the resin has fused into a homogeneous layer on the inner mold walls, the unit moves to the cooling chamber. Here, a water spray bath is used to cool and solidify the resin.

5. Moving to the unloading area, the molds are opened and the products removed. Product removal is a manual operation.

6. Flash is removed at the parting line and the parts are painted or finished as required.

Fig. 10-5. Large container is being removed from upper half of mold while mold on the right is being spray cooled. (McNeil Corp.)

Fig. 10-6. Multiple molds being charged with a predetermined amount of resin before the upper mold halves are clamped in place.

Rotational molding has a number of advantages over other molding systems. Total cost of equipment is comparatively less than other molding processes producing the same size moldings. It is possible to mold completely hollow products which is difficult with other processes. Extremely large items can be produced with relative ease. The major drawback to rotational molding is the long cycle period, which is continually being researched to speed up production.

APPLICATIONS

The versatility of design for rotational molding is almost unlimited. Flexibility or rigidity is built into the product by the use of many resins now available on the market. Typical products include large commercial and industrial containers, traffic cone markers, sporting equipment such as footballs, golf carts, and helmets, and insulated ice chests, Fig. 10-7. Other rotational molded products are illustrated in Fig. 10-8.

Fig. 10-8. This group of rotationally molded products includes an automotive arm rest, fuel tank, decorative bottle, open containers, and swimming pool safety floats.

cess. Many powdered resins are available for compounding and coloring. Using laboratory rotational molding equipment, such as shown in Fig. 10-9, it is possible to produce moldings similar to those used in industry. The only significant difference is in manually moving the mold through the various stages of the process. Fig. 10-10 shows the mold rotating mechanism. Note how the mechanism is constructed.

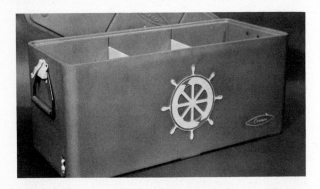

Fig. 10-7. An ice chest and lid both rotationally molded from polyethylene powder. Insulation is provided by the sealed double wall air pockets.

ROTATIONAL MOLDING IN LABORATORY

Many learning activities are provided through the study of rotational molding in the plastics laboratory. Mold design, construction, testing and production lead to a good understanding of the variations in the pro-

Fig. 10-9. A laboratory rotational molder equipped with temperature controls, timer, and rotational speed adjustment.

Fig. 10-10. The offset rotating arm holds the mold assembly while the fingers at the lower end of the mold clamp spin the mold on the secondary axis as they contact lugs within the oven during each revolution.

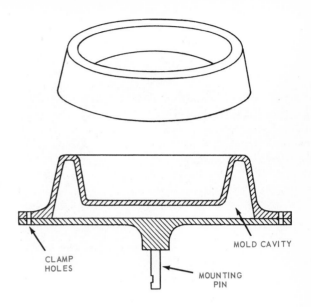

Fig. 10-11. Cross section of a cast aluminum rotational mold for a desk tray. A perspective view of the tray above.

The most common plastic powders for laboratory molding are both high and low density polyethylene, polyvinyl chloride, and powdered high impact polystyrene. Inert powdered colors are available for compounding almost any desired shade. Experimentation with fillers such as powdered metals, wood flour, and granular resins added to the molding powder will extend many desirable properties to the product.

LABORATORY MOLDS

Molds for laboratory production or experimentation are best made from cast aluminum or welded steel sheet. More intricate mold designs are possible with cast aluminum molds, while steel fabricated molds may provide easier construction for relatively simple product shapes. Fig. 10-11 illustrates a typical design for a cast aluminum mold. A cast mold requires the construction of a wood pattern and casting in the green sand foundry process. Machining and polishing are necessary on the inside mold surfaces to provide the desired surface texture on the product. A smoothly machined parting line will limit flash.

MOLDING SEQUENCE

The procedure for rotational molding should be carefully studied before actual molding. Specifications for the amount of powdered resin required for the product, oven temperature, and fusion time should be noted. The molding oven is then turned on and the heat regulator set for the desired temperature. For molding a round ball, Figs. 10-12 to 10-16, the resin charge can be prepared while the oven is heating. Both halves of the mold should be given an even coating of mold release, Fig. 10-12. Three quarters of a cup of low density polyethylene powder is emptied into a glass jar and one-forth of a tablespoon of powdered color added. The jar should be agitated until the color is well mixed with the powdered resin and then poured into the bottom half of the mold, Fig. 10-13. The mold halves are placed together and the wing nuts tightened securely. When the oven has reached 550 deg. F., the door is opened and the mold locked in the offset arm, Fig. 10-14, as quickly as possible to prevent much temperature drop. Protective or leather gloves should be used to handle all parts when the oven is hot. The timing control is set for 8 min. and rotational speed

Fig. 10-12. Mold release being sprayed on the interior halves of the rotational ball mold.

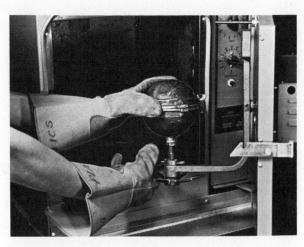

Fig. 10-14. Securing the charged mold to the offset arm of the rotational molder. Heat resistant gloves must be worn to prevent burns.

Fig. 10-13. The correct measure of powdered polyethylene is poured into the bottom half of the mold.

Fig. 10-15. Removing the solidified ball from the bottom half of the mold.

Fig. 10-16. The flash at the parting line is carefully trimmed with a sharp knife. Completed moldings are shown in the foreground.

at 20 rpm. Upon completion of the fusion cycle, the mold is removed with safety gloves and immersed in cold water until it is cool enough to handle with bare hands. The mold is then unclamped and the completed product removed as shown in Fig. 10-15. In most cases there will be a slight flash at the mold parting line which can easily be trimmed smooth with a sharp knife, Fig. 10-16.

STATIC MOLDING

Another process involving the use of powdered plastics is known as static molding. It is a minor process in the industry but deserves mention for it provides a way that huge containers and other products of extreme size can be produced.

The process consists of filling an open ended mold, usually made of sheet aluminum or steel, with powdered polyethylene as shown in Fig. 10-17. An insulating cover is placed on the mold and it is heated in an oven at 500 deg. F. or above. As the heat penetrates the mold, the resin next to the mold wall fuses and forms a skin. Varying wall thickness of the plastic can be controlled by the time the mold is allowed to remain in the oven. Wall thickness ranges from 1/4 to 1/2 in. for large products. When the desired wall buildup is obtained, the mold is removed from the oven and the remaining loose powder is dumped out. At this stage the inner wall is rough, and the mold is usually returned to the oven without the insulating lid for final curing and smoothing of the inside. The mold is then cooled with a water spray and the product removed.

The cycle is time consuming and limited to the production of only a few products per hour. Items such as chemical storage tanks measuring many feet in diameter and height are produced by static molding.

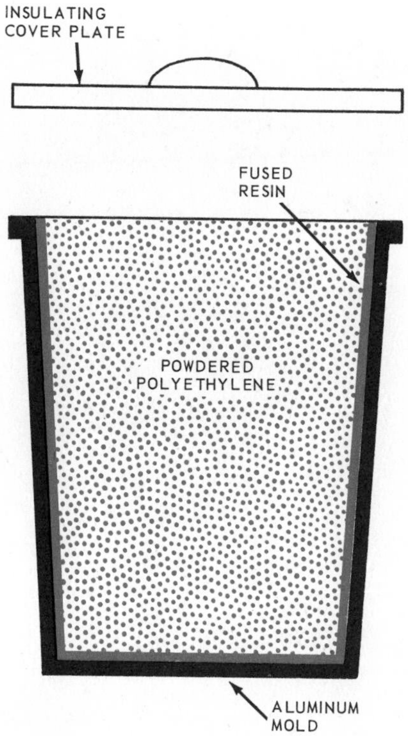

SPRAY COATING

A number of polymers, especially tetrafluoroethylene and others of the fluorocarbon family, are used in powder form for spray coating, Fig. 10-18.

When the dispersion spray method is used, the part must first be thoroughly cleaned. Areas not to be coated should be masked.

Fig. 10-17. Schematic cross section of a static mold as it would appear in the heating oven. The insulating cover keeps the resin from fusing across the top of the container.

Fig. 10-18. The non-stick properties of tetrafluoroethylene finished have led to widespread use of housewares. (Chemplast Inc.)

Surfaces to be coated are generally sandblasted to increase the bond strength of the coating. The part is then sprayed with a mixture of the powdered polymer in a liquid carrier to the determined thickness, Fig. 10-19. The coating is air dried and then heat cured in an oven. This fuses the individual particles of polymer to form a tightly bonded layer. The spraying, drying and curing cycles are repeated until the required coating thickness is achieved.

An electrostatic spray method can also be used. In this case, charged powder is sprayed on the part and then heated for fusion.

FLUIDIZED BED COATING

The process of coating products with powder in a fluidized state was developed in the early 1950's. It has now become an important coating method of the plastics industry, Fig. 10-20. The principle of fluidized bed coating is relatively simple. It consists of activating fine, solid particles of a plastic resin in such a way that they become suspended in air. Fig. 10-21 shows how a typical fluidized bed system operates. A powdered resin is placed in a tank which has a porous plate at

Fig. 10-19. Note dispersion spray coating of saw blades with tetrafluoroethylene. (Chemplast Inc.)

the bottom. Below the porous plate is a chamber where compressed air, at low pres-

Fig. 10-20. A large industrial fluidized bed coating system. The product moves from the liquid primer dip tank to the preheat oven, is placed in the fluidized dip, and is then cured in the post-heat oven. (Michigan Oven Co.)

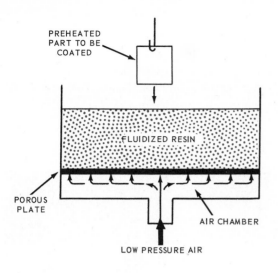

Fig. 10-21. Schematic drawing of a fluidized bed coating unit.
(W.S. Rockwell Co.)

sure, is introduced. When the air is turned on, it passes through the minute openings in the plate and causes the fine resin particles to actively float around in the tank. Too much air pressure would cause the particles to be blown from the chamber. When the correct amount of air is used, the resin particles float around the tank giving the appearance of a fluid or liquid. The openings in the porous plate must be small enough to prevent the resin particles from falling through. Fig. 10-21A shows the layout of a typical fluidizing bed coating facility.

Both thermoplastic and thermosetting resins can be used in the fluidized bed process to coat products with many desirable surfaces. Leaders among the powdered resins used in the process are polyvinyl chloride, epoxy, nylon, polyethylene, and the cellulosics. This process, which is used only for coatings, provides improved wear and corrosion resistance, surface smoothness, improved electrical properties, and a variety of colors for the product. See Fig. 10-22. Most coatings are applied to metal objects such as aluminum, brass, and steel.

Typical products being coated by the fluidized bed process include tool handles, electrical covers and transformer parts, wire drying racks, window sash, and steel pipe.

THE COATING PROCESS

The part to be coated, Fig. 10-22A, is usually dipped in a liquid primer, preheated and dipped into the bed of fine fluidized powder, Fig. 10-22B. The powdered resin particles melt and fuse on the heated metal surface to form a continuous uniform coating.

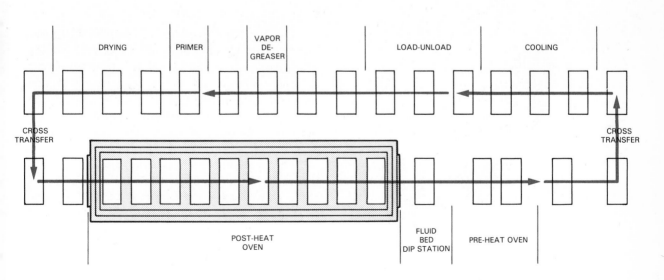

Fig. 10-21A. Floor plan layout shows a fluidized bed coating system. (W.S. Rockwell Co.)

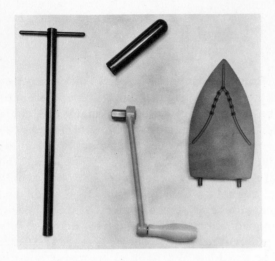

Fig. 10-22. A variety of metal products coated with plastic by the fluidized bed process. (W.S. Rockwell Co.)

The parts to be coated must be preheated to a temperature above the melting point of the powder to be used. Preheating temperature depends on the thickness of the part, desired coating thickness, and the type of plastic to be

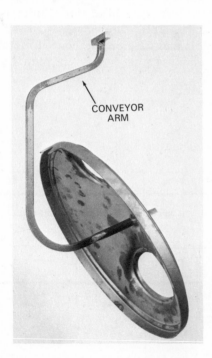

Fig. 10-22A. This transformer cover is hanging on the conveyor line at the loading station. (Michigan Oven Co.)

used, Fig. 10-23. After part is coated, it is returned to the oven for final curing and smoothing of the surface. In the case of thermosetting powders, the post-heat period polymerizes the resin. After removal from the oven, the product is water-cooled or allowed to cool in the air.

Fig. 10-22B. In-plant photograph shows part being dipped into the powder of a fluidized bed. (Chemplast Inc.)

FLUIDIZED BED COATING IN THE LABORATORY

The process used in industry for fluidized bed coating can also be performed in the plastics laboratory. A small industrial unit or one constructed in the laboratory will be quite satisfactory. Fig. 10-24 shows a fluidized bed coating unit constructed for laboratory use. Details are shown in Fig. 10-25.

The process consists of preheating the metal part in an oven at 350 deg. F. for 10 min. In this case a valve tool handle is to be coated with natural polyethylene. A wire must be attached to the part to hang it in the oven and also to hold it while dipping. While the part is preheating, air is turned on the fluidized bed unit so the powder rises to the top of the tank but does not overflow. The

Fig. 10-23. These transformer covers are on their way to the preheating oven from the prime dip tank.
(Michigan Oven Co.)

part is removed from the oven and quickly dipped into the fluidized powder with a continuous motion until particles no longer cling to the surface, Fig. 10-26. Return of the part to the oven for post curing, Fig. 10-27, should be done immediately. The curing phase takes about 3 min. at 350 deg. F. Upon final curing and cooling, the part is ready for use, Fig. 10-28, as no finishing is required.

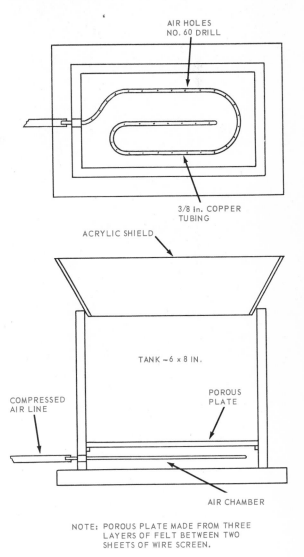

Fig. 10-24. A fluidized bed coating unit constructed in the plastics laboratory.

Fig. 10-25. Schematic details for the construction of a fluidized bed coating unit.

Fig. 10-26. Dipping the heated tool handle into the fluidized bed tank.

Fig. 10-27. The tool handle is placed in the oven for final curing. Most of the particles have already fused.

Fig. 10-28. The completed tool handle coated with polyethylene.

TEST YOUR KNOWLEDGE – CHAPTER 10

1. Rotational molding makes use of both _____ and _____ resins.
2. Products of extremely large size are produced by _____ molding and _____ molding.
3. The main resins used in powder molding and coating processes are:
   a._____.
   b._____.
   c._____.
   d._____.
4. List three typical products made by rotational molding.
   a._____.
   b._____.
   c._____.
5. Advantages of rotational molding over other processes include:
   a._____.
   b._____.
   c._____.
6. Rotational molds are usually made of _____.
7. The three main process requirements for a rotational molding system are:
   a._____.
   b._____.
   c._____.
8. Explain the rotating principle involved in rotational molding.
9. Rotationally molded hollow products may have open ends or areas by:

a._____.

b._____.

10. The_____ _____ _____ is the major drawback of rotational molding.

11. Static molds are usually made of_____ or_____.

12. In static molding, product wall thickness depends upon_____ and _____.

13. Explain the principle of fluidizing plastic powder in a tank.

14. Fluidized bed coatings are applied to materials such as:

a._____.

b._____.

c._____.

15. Three products coated by the fluidized

bed process are:

a._____.

b._____.

c._____.

16. Fluidized bed coating makes use of both _____ and_____ resins, adding desirable properties to products.

17. Preheating temperature of the part to be coated by the fluidized bed process is determined by:

a._____.

b._____.

c._____.

18. The two methods used for spray coating parts with powdered plastics are:

a._____.

b._____.

SUGGESTED ACTIVITIES

1. Make a product by the rotational molding process using a mold from your laboratory.

2. Write to manufacturers of powdered plastic resins and ask for specification sheets on the materials they produce. Prepare a bulletin board display depicting industrial powder molding from the materials you receive.

3. Prepare a procedure sheet for use with your laboratory rotational molder for a particular mold. Indicate volume of powder charge, oven temperature, rotational speed, and curing time.

4. Construct a fluidized bed coating unit for your laboratory. Select a product to be coated and perform the complete coating process.

5. Secure plastic powders of various resins and make a display illustrating the properties of each resin and the products for which they are used.

6. Write a paper on the development of powder molding and coating in the plastics industry.

7. Design and construct a mold for the static molding process. Test the mold by processing a few products.

8. Make a safety poster on the handling of hot molds in the rotational molding oven.

9. Set up an experiment to determine which mold releases are best suited for rotational molding of various powdered resins. Include the ease with which the product is removed from the mold, whether the application is practical, and surface finish on the product.

10. Make a chart indicating color samples available and amounts of color to be added to the powder charge for the molds for your laboratory rotational molding machine.

11. Contact local industries to discover the types of powdered resins they are using in their molding and coating processes. Secure samples if available.

12. Make a collection of pictures of products which have been powder molded or coated. Use these to make a bulletin board display.

Note automatic continuous thermoforming of small containers from sheet material.
(Gloucester Engineering Co., Inc.)

Chapter 11
THERMOFORMING
SHEET MATERIAL

Thermoforming plastic sheet material is one of the major processing techniques of the plastics industry and also one of the oldest. Since early attempts, about the turn of the century, to shape cellulosic sheet, the process has grown rapidly. Much of the growth is due to innovative techniques in forming equipment and to the development of new sheet materials with special thermoforming properties. See Fig. 11-1.

Fig. 11-1. Here is an assortment of thermoformed products including luggage, a portable cooler and furniture parts. (Brown Machine Co.)

Thermoforming is the heating of sheet plastic to a pliable state and forcing it around the contours of a mold by using pressure. The required pressure is usually mechanical or air pressure, often assisted by localized heating and bending. There are three main classifications of thermoforming with numerous variations within each category: Matched Mold Forming, Vacuum Forming, and Pressure Forming.

A wide variety of sheet plastic is suitable for thermoforming. Selection depends upon the requirements of the product, such as clarity, weatherability, toughness, flexibility, and color. Properties of sheet material for the thermoforming processes are listed in the reference section, page 271. Selection also depends on the process involved, as some sheet materials form better in one process than they do in another.

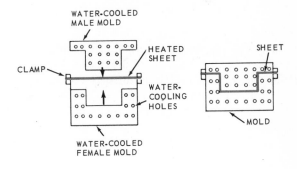

Fig. 11-2. Matched mold pressure forming is a highly positive thermoforming technique. (Mobay Chemical Corp.)

MATCHED MOLD FORMING

One of the more exacting techniques of thermoforming plastic sheet material, matched mold forming, requires two mold halves that fit together perfectly. The ther-

Fig. 11-3. Three matched mold thermoformers with molds for the production of refrigerator liners from sheet plastic.   (Hartig Plastics Machinery)

moplastic material is heated to its softening point and formed by mechanical pressure between the two halves of the mold. Fig. 11-2 illustrates a simplified method of the process. Since both halves of the mold contact each surface of the softened sheet, it is necessary that the mold surfaces be highly polished or textured to product requirements.

The molds are usually made of aluminum or steel and mounted in a hydraulic or pneumatic press. Heated sheet material is placed between the molds and the press is closed. The mold is normally water-cooled to control mold temperatures. Matched mold forming is especially suitable to products requiring excellent reproduction detail, Fig. 11-3.

VACUUM FORMING

Of all of the thermoforming processes, vacuum forming is the most versatile. It consists of heating a plastic sheet, held in a frame until it is soft and pliable. The mold is placed directly under the sheet and slight pressure is applied to seal the plastic to the upper mold edge. A vacuum is applied through small holes in the mold cavity and the atmospheric pressure forces the softened

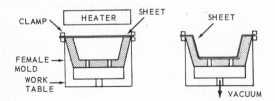

Fig. 11-4. Straight vacuum forming in a female mold is recommended for low-profile parts where deep draw is not a requirement.

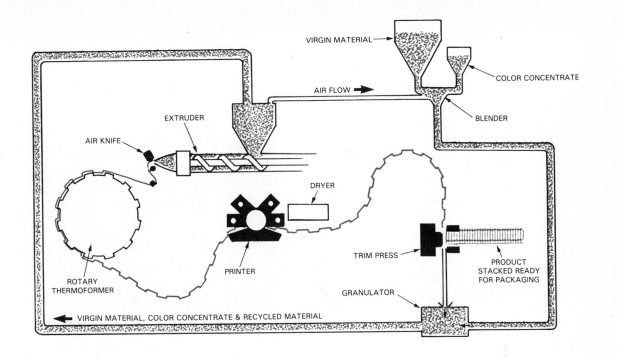

Fig. 11-5. Study the components of a continuous vacuum forming production system.
(Brown Machine Co.)

Fig. 11-6. This thermoforming machine is designed for volume production. It has three stations
for forming large parts from heavy gage material.   (Brown Machine Co.)

sheet against the contour of the mold walls, Fig. 11-4. Upon cooling, the product has solidified and when removed it retains the shape of the mold.

Vacuum forming is generally the least expensive of the thermoforming processes because the mold is made of one piece and is of relatively simple construction. Molds can be made of cast aluminum, machined aluminum, cast filled epoxy, wood, or plaster. Mold material selection is usually based on the production run. Hardwoods and plaster are ideal for prototype work and short production runs. Metal molds will produce large quantities of products with but little wear.

Production vacuum forming systems often include the extrusion of sheet material, vacuum forming, printing and trimming, all in one continuous operation. See Fig. 11-5. This is much faster and less expensive than performing all of these operations separately.

VACUUM FORMING MACHINES

Equipment for vacuum forming is manufactured in a variety of sizes and types. Extremely large products are formed on equipment mounted on the floor of the plant, while smaller items can be formed on relatively inexpensive machines. See Figs. 11-6 and 11-7.

The main requirements of any vacuum forming equipment include a heating device, clamping fixtures, mechanical assist, and a vacuum system. Air pressure for certain forming systems is also required. Most vacuum forming machines use infrared heaters which have variable temperature controls and are adjustable as to the distance from the forming sheet. Some machines are equipped with dual heaters above and below the sheet to provide for even heating of thick material.

A clamping frame is required in all vacuum forming processes to hold the sheet tight during heating, forming, and cooling. The frame also supports the sheet as it seals against the mold so as not to allow an escape of the vacuum. A mechanical assist, Fig.

11-8, is often used with the clamping frame to provide for even wall thickness on deep draw forming.

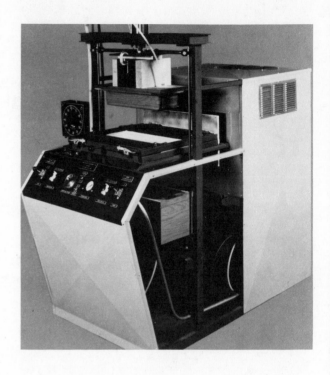

Fig. 11-7. This small manually operated vacuum forming machine is used for laboratory experimentation and short run production.   (AAA Plastics Equipment, Inc.)

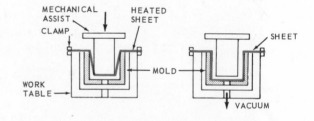

Fig. 11-8. Thinning of material in deep mold cavities can be overcome by use of a mechanical plug assist. The vacuum is drawn after the plug has stretched the sheet part way into the cavity.   (Mobay Chemical Corp.)

When the plastic has been softened prior to forming, the air within the cavity must be removed. This requires a pump and a large vacuum surge tank capable of recovery of the vacuum between cycles. The system should deliver a vacuum of about 28 in. of mercury.

Another process, using different machine types, is used in the production of small thermoformed containers. The equipment consists of a compression station, forming station and ejection station all in one machine, Fig. 11-8A. At the compression station, an extruder meters a precise amount of molten resin into a disc mold. The mold closes and forms a thin molten disc of resin ready for forming. The disc quickly moves to the forming station where it is drawn into a vacuum mold with a plug assist. The formed container then moves to the ejection station where it is removed from the machine and stacked. See Fig. 11-8B.

FORMING PROCESS

Many variations of straight vacuum forming have been developed to provide for a more even stretch of the material and a consistency of wall thickness on deep-drawn products. These include snap-back forming, Fig. 11-9, where a vacuum is drawn on the softened sheet to evenly stretch it down into the vacuum box. At the same time, a male mold is lowered into the cavity, the vacuum released, and the sheet quickly conforms to the contour of the male mold.

Fig. 11-8B. Top. Illustration shows a typical three-station forming machine. Bottom. A close-up view of forming and ejection operation.   (Hayssen Manufacturing Co.)

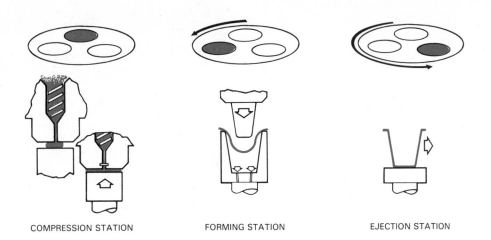

COMPRESSION STATION   FORMING STATION   EJECTION STATION

Fig. 11-8A. At the compression station, molten disc is being produced. A container is being formed at forming station and removed at the ejection station.   (Hayssen Manufacturing Co.)

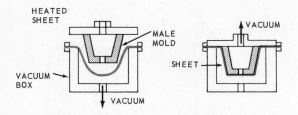

Fig. 11-9. Vacuum forming with snap-back can reduce starting sheet size, aid material distribution, and minimize chill marks.

illustrate other variations of vacuum forming for specialized processing.

Following the sequence through the vacuum forming process, the selected sheet material is cut to size and placed in the clamping frame, Fig. 11-13. The sheet is heated to the correct forming temperature

Drape forming, Fig. 11-10, makes use of a male mold and mechanical movement of the clamping frame. When the sheet is softened, the clamping frame is lowered over the mold which stretches the sheet to the general mold shape. A vacuum is then drawn, forcing the sheet tightly around the contour of the mold. Figs. 11-11 and 11-12

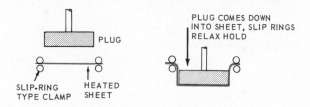

Fig. 11-12. Vacuum forming with plug and slip-ring produces strong parts with heavy wall sections in deep-draw setups.

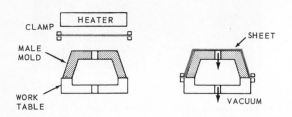

Fig. 11-10. Drape forming over a male mold usually results in better material distribution and depth-to-diameter draw ratios.

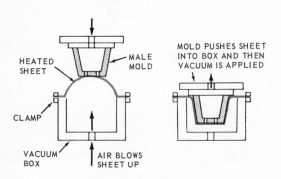

Fig. 11-11. Forming with billow snap-back is recommended for parts requiring a uniform, controllable wall thickness. (Mobay Chemical Corp.)

Fig. 11-13. Plastic sheet is located in the clamping frame ready for heating. The lowered male mold is shown at the bottom.   (The B.F. Goodrich Co.)

Fig. 11-14. The mold is raised into forming position and the vacuum drawn.

and the mold is brought into position for the vacuum draw as shown in Fig. 11-14. A short cooling period allows the sheet to solidify and the product is removed from the mold, Fig. 11-15. The product is then trimmed from the surrounding sheet by stamping, slitting, or sawing. Fig. 11-16 shows the finished product after trimming. A typical setup for trimming vacuum formed objects is shown in Fig. 11-17.

Fig. 11-16. After trimming, the completed shutter is ready for any necessary decorating.

The versatility of vacuum forming makes it suitable for extremely large products, as shown in Fig. 11-18, and high production of smaller items. Blister packaging, for example, Fig. 11-19, can be performed automatically on vacuum forming equipment as

Fig. 11-15. The frame has been unclamped and the formed sheet is being removed from the mold.

Fig. 11-17. A typical trimming setup for vacuum formed objects. Here a slitting saw trims the part which is guided by rollers and held on a trimming fixture.   (Cessna Aircraft Co.)

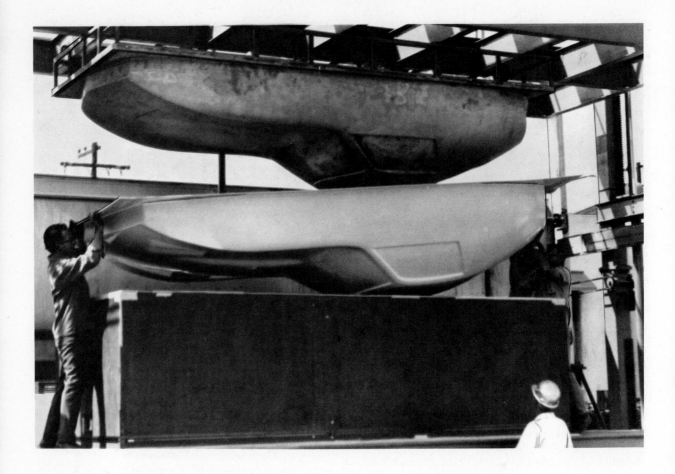

Fig. 11-18. Sheets of ABS-polycarbonate alloy 1/4 in. thick are used to vacuum form this complete automobile body. A body can be formed in a 20 min. cycle. (Borg-Warner Chemicals)

illustrated in Fig. 11-20. Food packaging takes advantage of the insulating qualities of foamed plastic sheet by vacuum forming containers. A few examples of vacuum formed food containers are in Fig. 11-21.

PRESSURE FORMING

The third main process in thermoforming is known as pressure forming, sometimes called air blowing. The process should not be confused with blow molding since it deals only with the forming of sheet plastic. The two primary techniques are Straight Pressure Forming and Free Blowing.

STRAIGHT PRESSURE FORMING involves a female mold over which a sheet of thermoplastic material is clamped, as shown

in Fig. 11-22. A radiant heater softens the sheet, a cover is quickly placed over the hot sheet, and preheated compressed air is blown through the cover opening. The sheet

Fig. 11-19. Vacuum formed blister packages are sealed to a cardboard backing. (Celanese Plastics Co.)

Fig. 11-20. An automatic vacuum former for forming, filling, and sealing containers and blister packages. (Brown Machinery Co.)

Fig. 11-21. Extruded foam polystyrene sheet is vacuum formed into a variety of packaging containers. (NRM Corp.)

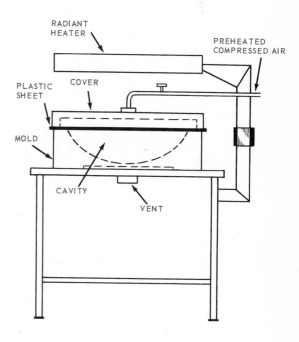

Fig. 11-22. Details of a sheet pressure forming machine.

is forced against the contours of the mold and any air trapped below the sheet escapes through vent holes in the mold. After cooling, the formed part is removed from the mold and is trimmed similar to vacuum forming.

FREE BLOWING is the process of placing a heated sheet of plastic over a pressure box and clamping a frame with a given opening in it over the sheet. The opening may be circular, square, oval or other required shape. Compressed air is forced from the pressure box causing the sheet to form a smooth bubble, Fig. 11-23. Heat and pressure provide for the desired height of the bubble. Acrylic sheet is commonly used in this process as it provides good optical clarity for such products as building panels and canopies for aircraft. The softened plastic expands like a balloon.

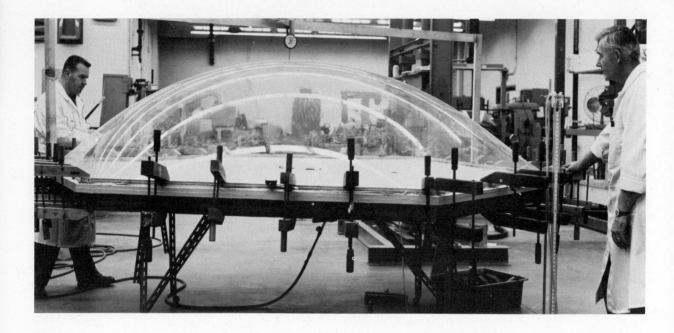

Fig. 11-23. Acrylic sheet being free blown in a clamping frame. An indicator at the top shuts off the air pressure as the sheet reaches the correct height.   (Rohm and Haas Co.)

THERMOFORMING IN THE LABORATORY

Many of the industrial techniques of thermoforming sheet material can successfully be accomplished in the plastics laboratory. The student has an opportunity to become better acquainted with the properties of plastic sheet, the principles of mold design and construction, and sheet forming equipment as he duplicates the procedures followed by industry.

LABORATORY PRESSURE FORMING

A variety of product and mold designs for pressure forming should be sketched and the most suitable selected for construction. Wood molds are quite adequate for the forming process. The following sequence provides a typical procedure for the sheet pressure forming process.

A sheet of high impact polystyrene is cut to the correct size and placed over the mold cavity, Fig. 11-24. The sheet may be heated in an oven, on a hot plate, or with a heat gun as shown in Fig. 11-25. Since measuring the

temperature of the sheet is difficult, by touching the sheet with the soft eraser of a lead pencil the correct forming temperature can be determined. The sheet should not sag but be slightly rubbery to the touch of the pencil. The mold cover is quickly closed and held firmly while compressed air is forced into the cover hole at a pressure of 20 to 40 psi, Fig. 11-26. Air pressure should be held for a few seconds to cool the sheet, the cover opened, and the product removed as shown in Fig. 11-27. The container is then ready for trimming with shears or on a fixture mounted in a lathe as shown in the section on vacuum forming.

LABORATORY VACUUM FORMING

The plastics laboratory provides an excellent opportunity for the study of vacuum forming techniques used in industry. Laboratory size vacuum formers range from semiautomatic units, Fig. 11-28, to manually operated machines, Fig. 11-29. Each is adaptable to a variety of mold sizes and shapes allowing for an unlimited number of product design possibilities. The com-

Fig. 11-24. This pressure forming mold has a hinged cover with narrow strips of band saw blade placed along the edge to firmly grip the hot plastic sheet.

Fig. 11-26. A blast of compressed air forces the soft sheet against the walls of the mold cavity.

plete forming process involves the design of a product, mold construction, sheet material selection, a study of the vacuum form-ing cycle, and the necessary trimming system. The process is illustrated by following through the procedure of making two vacuum formed parts.

The design problem requires a lid for a container. The container is the lower half of a one-gallon bleach bottle made of high density polyethylene and readily obtainable. A wood fixture is constructed to mount on a

Fig. 11-25. Softening the plastic sheet to forming temperature with a hot air heat gun.

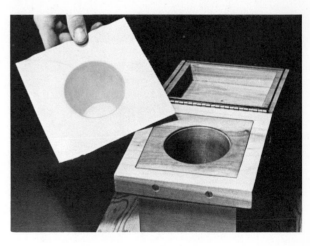

Fig. 11-27. The pressure formed container is removed from the mold and is ready for trimming.

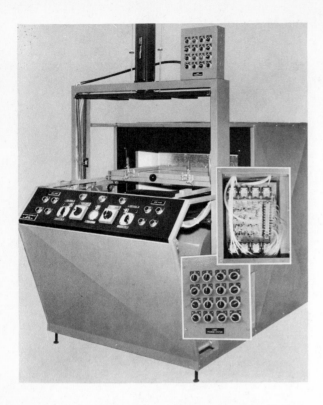

Fig. 11-28. A semiautomatic vacuum forming machine featuring controlled heating and forming cycles. (AAA Plastic Equipment, Inc.)

Fig. 11-30. Removing the top half of a plastic bleach bottle on the lathe. The bottle is held in the wood fixture by friction as the knife slices through the thin plastic.

standard wood lathe to cut off the top half of the bottle as shown in Fig. 11-30. The necessary measurements are then made to determine the shape and size requirements for a suitable lid. Fig. 11-31 shows a cross section

Fig. 11-29. A laboratory vacuum former with manually operated clamping frame, radiant heater, and platen.

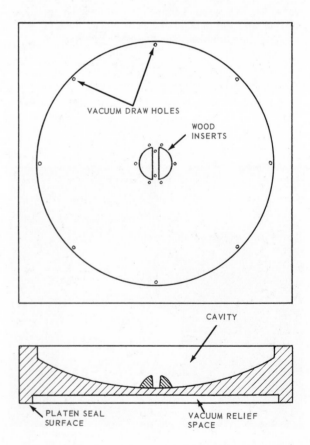

Fig. 11-31. Cross section of a hardwood mold for a vacuum formed container lid. Wood inserts provide handle shape.

of a mold design selected for the container lid. The mold is made of hardwood and the cavity section turned and sanded smooth on the lathe. A small split turning is glued to the center of the cavity to produce the handle grip on the plastic lid. Small holes are drilled at the spots requiring the deepest draw, and the bottom of the mold is recessed to provide air space for the vacuum, Fig. 11-32. Forming is done in the vacuum unit and the product is removed, Fig. 11-33, ready for trimming.

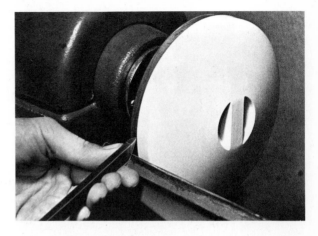

Fig. 11-34. Removing excess material on a trimming fixture in a lathe.

Fig. 11-32. Completed wood mold for the container lid with plastic sheet cut to size ready for forming.

The completed lid is then tested on the container for final fit, Fig. 11-35.

Basically the vacuum forming process is as follows:

1. The mold is selected from the laboratory, Fig. 11-36, or constructed by the student. Fig. 11-37 shows drawing details of the mold used in this procedure.

2. The clamping frame is adjusted to fit the mold size, the mold is placed on the platen, and the platen is raised to position, Fig. 11-38.

A trimming fixture is made from a piece of hardwood turned on the lathe providing a pressure fit for the lid. A skew chisel is used to remove the excess sheet and smooth the edge of the lid as shown in Fig. 11-34.

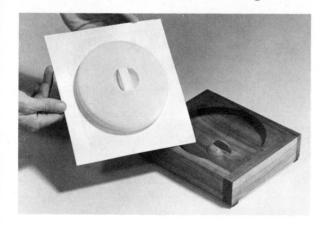

Fig. 11-33. The formed container lid is inspected after removal from the mold.

Fig. 11-35. Checking the fit of the vacuum formed container lid.

Fig. 11-36. A group of vacuum forming molds ready for use.

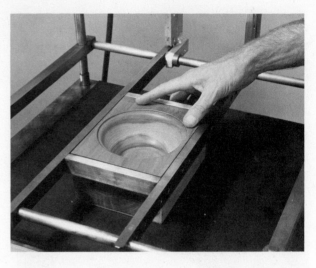

Fig. 11-38. Centering the mold in the lower half of the clamping frame. It should be flush with the surface of the frame.

3. A sheet of high impact polystyrene .060 thick, 6 x 8 in., is placed over the mold, Fig. 11-39, lapping over the edges of the lower frame.

4. The hinged upper frame is lowered over the sheet material and locked in place, Fig. 11-40.

5. The vacuum pump is turned on to begin building up a vacuum ready for forming, while the radiant heater is turned on and swung into position over the plastic sheet, Fig. 11-41.

6. When the plastic has softened to a rubbery state, the vacuum switch is thrown, forcing the sheet to conform to the mold walls, Fig. 11-42.

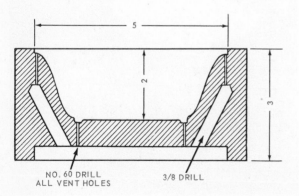

NO. 60 DRILL
ALL VENT HOLES

3/8 DRILL

5

2

3

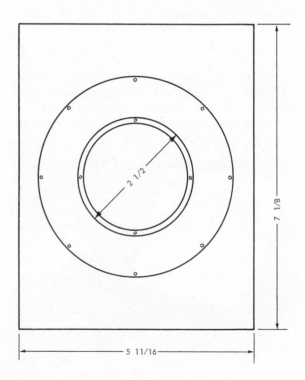

2 1/2

7 1/8

5 11/16

Fig. 11-37. Working drawing of a mold for a vacuum formed dish.

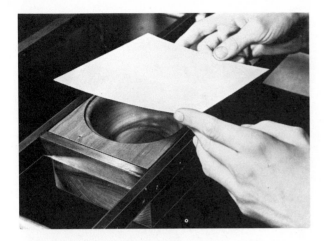

Fig. 11-39. Locating the styrene sheet over the mold.

Fig. 11-41. The radiant heater is swung over the mold table about 4 in. above the sheet plastic.

7. The vacuum is held on until the sheet is cool enough to retain its shape, the clamping frame released, and the part removed from the mold as shown in Fig. 11-43.

8. The part is placed on the trimming

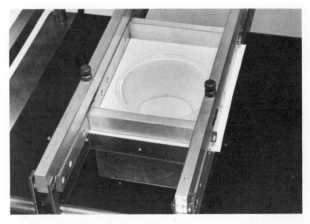

Fig. 11-42. Drawing down the softened sheet with the vacuum turned on.

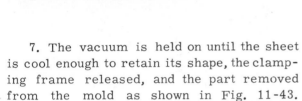

Fig. 11-40. Clamping the sheet between the two metal frames so it cannot slip when the vacuum is drawn.

fixture in the lathe and slit at the trim line with a skew chisel, Figs. 11-44 and 11-46.

9. The edge of the dish is lightly sanded and the dish completed, Fig. 11-45.

The trimming fixture, Fig. 11-46, is made after a trial dish has been formed. It is turned from hardwood to fit the contour of the inside of the dish. The lathe tailstock,

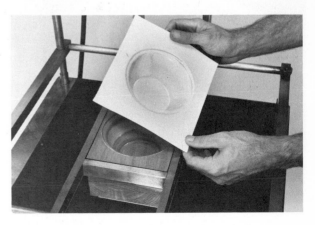

Fig. 11-43. Removing the formed part from the mold.

Fig. 11-44. Excess sheet has just been cut away from the formed dish on the trimming fixture.

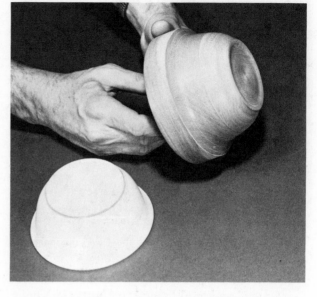

Fig. 11-46. The trimming fixture holds the formed part securely in place while excess sheet stock is cut away.

with a block of wood against the bottom of the dish, should press firmly against the fixture and be locked in place. Safety glasses should be worn when trimming. The parts involved in this total forming process are shown in Fig. 11-47.

SKIN AND BLISTER FORMING

The packaging industry makes use of a number of vacuum forming techniques known as skin and blister forming. Skin forming refers to the process of heating a thin sheet of plastic to its softening point, lowering it

over a product, and drawing a vacuum as in Fig. 11-48. No mold is involved. The sheet material conforms to the shape of the product appearing as a plastic skin. The product is then usually sealed between the plastic skin and a cardboard backing, Fig. 11-49, and is ready for distribution. Vacuum skin packaging provides protection for the product, makes inexpensive containers, provides clarity to view the product, and adapts well to high-speed production.

Fig. 11-45. A completed dish is filled with candy ready for serving.

Fig. 11-47. The wood vacuum mold, a formed sheet, the trimming, and the completed dish.

Fig. 11-48. A heated cellulose acetate sheet is ready to be drawn tightly around a C-clamp by the vacuum on this laboratory vacuum former.

Blister forming is very similar to skin forming except a mold is normally used to produce a given shape in the plastic sheet into which the product is placed. The formed shape in the sheet is called a blister. A blister forming mold is shown in Fig. 11-50. A plastic sheet is placed in the forming machine over the mold. Look at Fig. 11-51. The sheet is heated until soft, then lowered to the mold surface, and finally the vacuum is applied. Upon cooling, the sheet is removed from the machine. See Fig. 11-52. It is then trimmed to size with shears or a paper cutter.

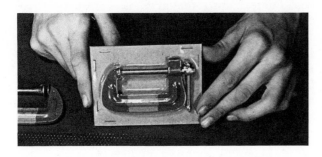

Fig. 11-49. The skin formed sheet is trimmed and fastened to a cardboard backing similar to industrial processes.

For this illustration, a laboratory display is being made by blister packaging a number of different resins as shown in Fig. 11-53. A sheet of clear plastic is then sealed over the blister formed sheet enclosing the gran-

ules in the individual cavities. Each resin is then labeled for identification.

HEATING AND BENDING PLASTIC SHEET

Although heating plastic sheet for mechanical bending is usually related to fabri-

Fig. 11-50. A blister forming mold made from turned wood blocks mounted on a Masonite backing ready for forming.

cation, it is a thermoforming process. It is not a major industrial process but is worthy of discussion. The process can be broken down into two areas: heating a section of a sheet for localized bending, and softening a complete sheet for more intricate forming. Most thermoplastic materials work satisfactorily for these processes.

Fig. 11-51. Sheet plastic being placed over the mold prior to softening by the radiant heater at the top.

Fig. 11-52. Removing the formed sheet from the clamping frame.

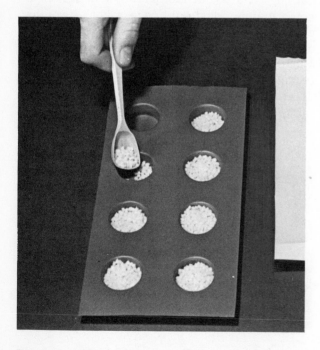

Fig. 11-53. Adding granules to the formed sheet for a packaging display.

A strip heater provides a good heat source for localized bending. The plastic sheet is marked for the bend and placed over the heater until the area is softened, Fig. 11-54. It is quickly moved to a form to hold it in exact position until it cools, as shown in Fig. 11-55. Wood fixtures are adequate for almost all bending applications. If a number of parts are required with the same bend, a fixture with stops and clamps can be constructed to meet the particular need.

Mechanical forming of plastic sheet requires that the whole sheet be heated to the forming temperature. The forming temperature for various types of plastic sheet can be found in the back reference section. Sheet material can be heated in an oven, placing it on a flat sheet of stainless steel, or other metal, which will keep it flat while softening. A heat gun or hot plate can be used for softening the material.

Thermoforming Sheet Material

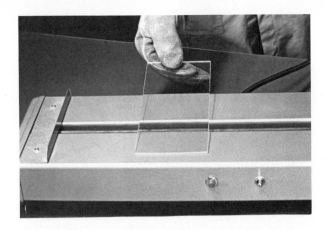

Fig. 11-54. Locating plastic sheet over the heating element of a strip heater.

Fig. 11-55. Holding the bent sheet against a wood form to retain the desired shape.

tion, an acrylic fixture is being made to preform veneer strips for laminating, Fig. 11-57. The plastic sheet is usually cut to the exact size and the edges smoothed and polished before forming.

An unlimited number of product design problems make use of thermoforming by heating and bending. Additional fabrication of sheet plastic is covered later.

Fig. 11-56. The softened sheet is held securely to a round wood form until cool.

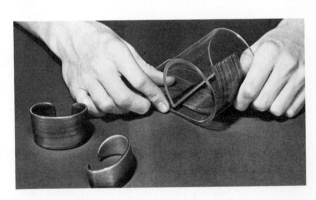

Fig. 11-57. Veneer strips, soaked in water, are pressed into the formed acrylic fixture to dry. The preformed strips will then be laminated into a bracelet as shown at the left.

When the sheet has reached the proper temperature, it is removed from the oven, with heat protective gloves, and bent to the shape of a form, Fig. 11-56. It must be held to the contour of the form until cool enough to retain the desired shape. In this illustra-

TEST YOUR KNOWLEDGE – CHAPTER 11

1. Give a general definition of thermoforming plastic sheet.
2. The three main categories of thermoforming are:
   a._____.
   b._____.
   c._____.
3. In matched mold forming _____ or _____ pressure is used to close the mold halves.

4. Compared to other sheet forming methods, matched molding is:
   a._____.
   b._____.
   c._____.
   d._____.
5. Vacuum forming molds can be made from:
   a._____.
   b._____.
   c._____.
   d._____.
   e._____.
6. Deep drawing of vacuum formed sheet is often unsatisfactory due to thin wall sections. This is overcome by using _____ forming or _____ forming.
7. Four advantages of vacuum skin packaging are:
   a._____.
   b._____.
   c._____.
   d._____.
8. How does blister forming differ from skin packaging?
9. A _____ _____ is often used for localized bending of plastic sheet.
10. Three methods of heating plastic sheet for mechanical forming are:
    a._____.
    b._____.
    c._____.
11. Matched mold forming provides:
    a. Outstanding scratch resistance.
    b. Little waste from scrap.
    c. Excellent reproduction detail.
    d. All of the above.
12. The most versatile of all thermoforming processes is:
    a. Vacuum forming.
    b. Pressure forming.
    c. Heating and bending.
    d. None of the above.
13. The main requirements of a commercial vacuum forming machine are:
    a._____.
    b._____.
    c._____.
    d._____.

SUGGESTED ACTIVITIES

1. Design and construct a mold for a laboratory vacuum forming machine. Test the mold by forming some products of high impact styrene or other sheet plastic.
2. Prepare a sheet of high impact polystyrene by compression molding of granules on a hydraulic press. Test the sheet by vacuum forming it in a mold.
3. Collect a variety of thermoformed products and make a display showing some examples of how thermoforming of sheet material is used in industry.
4. Design a product that requires heating and bending of plastic sheet and make the necessary fixtures to hold the sheet in place while cooling.
5. Write to manufacturers of thermoforming equipment and obtain pictures of their machines. Prepare a bulletin board showing types of machines used in industry.
6. Visit a thermoforming company in your locality and make a listing of the products they make and the thermoforming processes used.
7. Using equipment from your laboratory, demonstrate how pressure forming is done.
8. Plan an experiment which will illustrate the depth of draw and wall thinning for a variety of thicknesses of plastic sheet.
9. Collect inexpensive thermoformed products and do identification burning tests on them to see what resins were used.

Chapter 12
CASTING PLASTICS

Casting refers to any of a number of processes in which a liquid plastic is poured into a suitable mold. The plastic may be in the form of a monomer which is polymerized after it is poured, or a liquid resin to which a catalyst is added prior to pouring. The molds range from complicated to simple home workshop types. See Fig. 12-1. Resins

Fig. 12-1. Precision industrial parts cast from plastic resins. Note fine detail.
(General Electric Co.)

can be cast cold or hot and allowed to solidify through further polymerization with heat added if necessary.

Common casting resins are the acrylics, polyesters, nylons, silicones, epoxies, phenolics, and polyurethanes. Resins are often filled for reinforcement. They are available from clear to opaque and may be colored to any desired shade.

Molds used in casting processes are usually inexpensive and are made from materials such as plaster, glass, wood, metal, and other plastics. A major use of casting in the plastics industry is to produce cast acrylic sheet, tubing, and rods, Fig. 12-2. Phenolic billiard balls, jewelry, and imitation marble are made by casting. The polyesters are used for embedding scientific specimens, hobby work, cast sheets, furniture parts and jewelry. Epoxy casting resins commonly filled with metal powder are used extensively in the automotive and aircraft industries for making molds, tooling jigs, and dies. Polyurethane is used in both flexible and rigid casting for items such as furniture parts, automotive crash padding, and housewares. The flexibility of silicone resins has made them ideal for casting molds into which other resins are cast.

CASTING PROCESSES

One of the major casting processes, that of producing cast acrylic sheet, is done by using a premixed acrylic monomer in the form of a thick syrup. The liquid is poured between two sheets of polished glass and allowed to cure. After casting, the sheets are heated to relieve any stresses, cut to size, and paper coated to protect the polished surfaces. A process similar to this is used for casting acrylic rod, Figs. 12-2 and 12-2A.

Fig. 12-2. Crystal clear acrylic rods are made by casting an acrylic monomer syrup in glass molds. Here they are being placed in a heating chamber for curing. (Cadillac Plastic and Chemical Co.)

Fig. 12-2A. These are cast acrylic stock shapes and specialty items. (Glasflex Corp.)

Another production casting process makes use of polyester casting resins. The resin and catalyst are mixed automatically in a dispenser unit, as in Fig. 12-3, and the liquid polyester is ready for continuous casting, Fig. 12-4. Production furniture and cabinet parts, made by casting in this manner, closely resemble wood.

Fig. 12-3. A 55 gallon premix polyester dispensing unit. The resin is ready for continuous production casting. (Pyles Industries, Inc.)

Fig. 12-4. Polyester casting resin is poured into a mold in the production of furniture parts as shown on the right.

Nylon is an important casting material because of its high strength and wear resistance. The monomer is cast into one or two piece molds and cures into the polymer. Size and thickness are limited only by the size of the mold. Size is a major limitation when injection molding nylon. Mold costs are also low for cast parts, making short runs feasible. Typical parts include heavy duty bearings, gears, boat propellers and chemically resistant containers, Fig. 12-4A.

Silicone resins are used extensively for making molds into which other plastics may be cast. The properties of silicone as a mold material include flexibility for easy part removal and exacting reproduction of detail. No mold release is required for part ejection. Lead may even be cast into a silicone mold since it is resistant to temperatures up to 600 deg. F. (316 deg. C).

Fig. 12-4A. This is a heavy walled cast nylon container. (The Budd Co.)

The preparation of a silicone mold begins with the selection or construction of a pattern. The pattern is fastened to the base of a mold frame and coated with mold release, Fig. 12-5. The silicone resins and catalyst are weighed in proportional amounts, Fig. 12-6, and mixed together, Fig. 12-7. Bubbles caused by mixing are removed under a vacuum bell jar, Fig. 12-8. Then, the resin is carefully poured over the pattern to avoid trapping air bubbles, Fig. 12-9. When the mold frame has

Fig. 12-5. To insure easy removal of the pattern, it is coated with mold release. (General Electric Co.)

Fig. 12-6. Thick bodied silicone resin being weighed, a proportional amount of catalyst is weighed separately.

Fig. 12-7. Catalyst and resin are mixed. (General Electric Co.)

Fig. 12-8. The resin is being deaerated in a vacuum chamber.

been completely filled with silicone, a cover is placed over the mold, Fig. 12-10, to produce a flat surface. After a 24 hour curing period, the back plate is removed and the pattern is stripped from the mold, Fig. 12-11. The finished mold, Fig. 12-12, is now ready for casting with any number of different liquid resins. Parts that are cast in silicone molds may be pigmented internally or painted, Fig. 12-13. Silicone molds used for casting have an average life of 100 to 200 molding cycles.

Fig. 12-9. Silicone resin is poured into mold containing the pattern.

Fig. 12-11. The pattern is removed from the cured silicone. (General Electric Co.)

Fig. 12-12. The new silicone mold on the right is now ready for production.

Fig. 12-10. Backing board is removed after curing is completed.

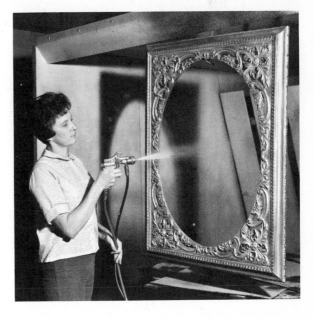

Fig. 12-13. Urethane picture frame is spray finished in antique gold. This intricate part was cast in a silicone mold.

Embedding objects in plastics is accomplished with a smooth surfaced mold and a clear acrylic or polyester resin. This is done in a two step casting process. First, a catalyzed layer of plastic is poured into the mold and allowed to cure. Then, the object to be embedded is placed on the partially cured surface. The second layer is poured into the mold, surrounding the object and bonding with the first layer. Plastic embedments are used for clear display of valuable items, preservation of biological specimens, and for functional parts as shown in Fig. 12-14.

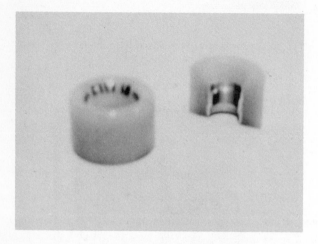

Fig. 12-14. Cast polyurethane roller skate wheel with metal bearing embedded.

Electronic components can be protected using a potting operation. The components are located in a housing and encased with plastic resin, Fig. 12-15. This provides protection from moisture, heat and impact.

CASTING IN LABORATORY

When casting in the plastics laboratory, thermosetting resins -- the polyesters, silicones and epoxies are generally used. Plastisols (mixtures of polyvinyl chloride and liquid curing agents) may also be cast. Plastisol molding is discussed as a separate topic in the next chapter.

Most casting materials consist of a resin base to which a catalyst is added to cause final polymerization or hardening. The process involves:
1. Preparation of a suitable mold.
2. Pouring the resin.
3. Removal of product from the mold.

Molds can be made of many materials including wood, glass, polyethylene, silicone and a number of metals. A mold release is generally used to permit easy removal of the product.

Polyester and epoxy casting resins are quite similar. They are obtained as a clear resin base to which a clear catalyst is added. The casting procedure is as follows:
1. Prepare or obtain a suitable mold and apply a thin coating of mold release to the inside surfaces.
2. Pour into a measuring cup the correct

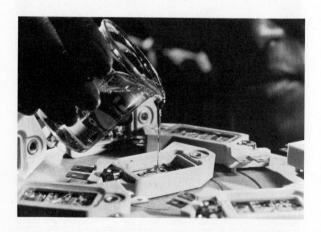

Fig. 12-15. For protection, a clear flexible silicone plastic is cast over electronic connections in computer circuit. (General Electric Co.)

Fig. 12-16. Catalyst is added to the polyester resin along with the desired color before mixing.

amount of resin to completely fill the mold and add the amount of catalyst as specified by the manufacturer, Fig. 12-16.

3. The mixture should be stirred slowly for a few minutes being careful not to trap air bubbles. If color is desired, liquid color concentrate, as required, is added.

4. The resin mixture is then poured slowly into the mold to the correct level as shown in Fig. 12-17. Inexpensive glass molds in various sizes are available from plastics supply houses.

5. Upon curing, which takes from 2 to 6 hours depending upon the amount of catalyst used, the product is removed from the mold, Fig. 12-18.

6. Final finishing can be accomplished by filing and buffing with a polishing wheel.

The general procedure described is used with all casting resins and molds with slight variations. The silicone resins are useful in the laboratory for casting molds for short production or intricate cast plastic products. The process involves two casting sequences. A pattern made of wood, plastic, metal or other material is placed on a mold plate. A small amount of contact cement is used to keep the pattern in place. The mold frame is placed around the pattern with paste wax on the edges to keep the poured resin from leaking out. The amount of catalyst specified by the manufacturer is added to the silicone resin and the mixture is stirred slowly. Slow mixing will avoid trapping air bubbles in the resin. The silicone resin is slowly poured into the mold starting at one edge, allowing the resin to cover the pattern until the mold is full, Fig. 12-19. When the silicone mold has cured, the pattern is removed, Fig. 12-20.

Fig. 12-17. Pour the polyester resin slowly into a glass mold. The project being cast is a gear shift knob. Later it will be drilled and tapped.

Fig. 12-19. With the catalyst added, the silicone resin is poured over the pattern in the mold frame.

Fig. 12-18. The glass mold is broken by wrapping it in a cloth and tapping lightly with a hammer. Final polishing may be done with a soft cloth.

A casting resin is then prepared, as has previously been described, and poured into the silicone mold, Fig. 12-21, until the mold is full. The resin is allowed to cure and is removed from the mold in the same manner the pattern was removed. The edges of the casting are filed smooth, buffed, Fig. 12-22, the background painted, and the product completed, Fig. 12-23.

Fig. 12-20. A slight twisting of the flexible silicone mold easily releases the pattern.

Fig. 12-22. Buffing the edge of the painted casting for a final finish.

Fig. 12-21. Polyester casting resin being poured into the silicone mold.

Fig. 12-23. The completed polyester casting with a painted background to bring out the colored letters.

Castings with embedded specimens are made by pouring the resin to the desired level and gently placing the specimen on the resin surface. The remainder of the resin is then gently poured over the specimen being careful not to trap air bubbles. If the specimen is too heavy, the two layer casting process can be used.

TEST YOUR KNOWLEDGE – CHAPTER 12

1. Materials used for molds for casting plastics include:
   a. _____.
   b. _____.
   c. _____.
   d. _____.

2. Molds, tooling jigs, and dies are often cast from _____ resins.
3. Acrylic sheet is often cast in a mold made of _____.
4. Cast molds in which other materials are cast are usually made of:
   a. Polyester resin.
   b. Silicone resin.
   c. Polycarbonate resin.
   d. All of the above.
5. Casting resins are usually in the form of a _____ or a liquid to which a _____ is added.
6. Ideal resins for casting in the laboratory are _____, _____, and _____ .
7. _____ resins are used for embedding scientific specimens and hobby work. One example, a rare coin may be displayed in clear plastic.

SUGGESTED ACTIVITIES

1. Obtain or construct a mold and make a polyester casting.
2. Review the latest available literature and write a report on industrial processes used to produce cast plastic furniture.
3. Design a mold suitable for embedding a rare coin or other specimen in a casting resin and make the casting.
4. Prepare a display of cast plastic products. Indicate the type of resin and the casting process for each product.
5. Make a pattern and pour a silicone mold for an epoxy or polyester casting. Test the mold by making a cast product.
6. Write to companies that manufacture plastics for casting and request specifications on the resins they make. Prepare a paper from the material received on the present use of plastic casting in industry and predict future uses.

Automatic plastisol dip coating system can be used for coating parts like plier handles and wire cutters.   (W.S. Rockwell Co.)

Chapter 13
PLASTISOL MOLDING

Plastisols are a mixture of finely ground polyvinyl chloride resins and plasticizers (chemicals added to improve workability, and to reduce brittleness). At room temperature plastisols vary from water thin to thick syrup consistency. If the temperature is raised to 350 deg. F., fusion of the fine particles take place and the resulting mass is a tough, flexible, solid material. Plastisols may be converted from a liquid to a solid without pressure, and are adaptable to a variety of molding and coating processes using simple, inexpensive equipment.

Since plastisols are mostly solids, little loss in weight occurs during fusion. They can be poured or pumped into a mold, sprayed on the surface, or a mold may be dipped into the liquid. Plastisols are available in a wide range of colors and degrees of flexibility.

DIP MOLDING

Dip molding is the process of dipping a heated male mold into a liquid plastisol, fusing the plastic on the mold surface, and stripping the finished product off the mold. Typical products produced by the dip molding process are spark plug covers, toys, boots, eyeglass cases. The product may be used as the mold, and the plastisol coating retained. Typical products are tool handles, machine knobs, household dish drain racks, Fig. 13-1. Additional plastisol-coated products are shown in Fig. 13-2.

Fig. 13-2. Note typical dip coated products.

Fig. 13-1. Welded wire refrigerator rack is dip coated with polyvinyl chloride plastisol.   (The B.F. Goodrich Co.)

The industrial dip molding system usually includes a conveyor line passing through a preheating oven, dipping station, Fig. 13-3, fusing oven, cooling area, and stripping area. Normally the mold is heated to approximately

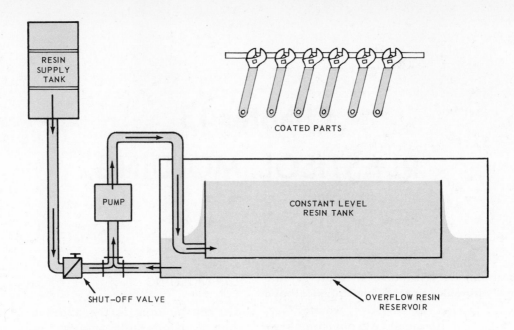

RESIN
SUPPLY
TANK

PUMP

CONSTANT LEVEL
RESIN TANK

COATED PARTS

SHUT—OFF VALVE

OVERFLOW RESIN
RESERVOIR

Fig. 13-3. Schematic of the dipping station of a conveyor line dip molding system.
(W.S. Rockwell Co.)

Fig. 13-4. Plating rack being withdrawn from the plastisol
tank is ready for oven curing.
(Michigan Chrome and Chemical Co.)

300 deg. F., providing a wall thickness of 1/16 to 1/8 in., before dipping. The mold is moved to the dipping station and is immersed in the liquid plastisol. It remains in the plastisol from 3 to 5 min. and is then withdrawn slowly, Fig. 13-4. The plastisol on the mold is then fused by placing it in an oven heated to about 350 deg. Fusing time varies from 5 to 15 min. Upon completion of the fusing, the mold is cooled by a water spray. The product is then stripped from the mold or, in the case of a coated product, the product is complete. Fig. 13-4A shows an automatic dip coating machine used for coating small products.

Cold dip plastisols are also available for coating parts that are too delicate to be heated. These are often called tool dip plastisols and only require air drying, Fig. 13-4B. Oven curing is not needed.

As the details of the mold are on the inside, the product must be turned inside out.

High production rates may be obtained in dip molding by using a conveyor system. A conveyor can dip a large quantity of parts in succession.

Plastisol Molding

Fig. 13-4A. This plastisol coating system shows coated bus bar boots on their way from the dip station to oven curing. (W.S. Rockwell Co.)

SPRAY COATING

Plastisols are used in spray coating systems for insulation, sound deadening, and protective coatings. Standard heavy-duty spray equipment may be used to apply the plastisol. Hot air is generally used for curing. By spraying, coatings up to .050 in. in thickness may be obtained.

Linings of railroad tank cars are often spray coated with polyvinyl chloride plastisols to provide chemical resistance, Figs. 13-5 and 13-6. Hot air is forced into the tank to provide fusion.

Fig. 13-5. Plastisol spray equipment is readied for coating inside of railroad tank car.

Fig. 13-4B. Cold dip plastisol on tool handles such as pliers provides electrical insulation. Curing in oven is not needed.

Fig. 13-6. Worker spraying the inside of a tank car prior to fusion of the plastisol.

SLUSH MOLDING

Slush molding is a form of molding which takes advantage of the ability of plastisols to solidify immediately on contact with heat. A hollow preheated mold, usually aluminum, is filled with liquid plastisol. The material at the wall of the mold begins to gel immediately upon contact. Plastisol is retained in the heated mold from 3 to 5 min., and the wall thickness continues to build up. The mold is dumped before the entire mass becomes solid, which drains off the remaining liquid plastisol. Returning the mold to the oven for a few minutes at 350 to 400 deg. F. fuses the inside plastisol wall. After water cooling, the mold is opened and the product is removed. Wall thickness is determined by the mold temperature and the dwell time (period of time plastisol remains in hot mold). Fig. 13-7 illustrates a continuous production slush molding system used in industry.

Plastisol slush molding is used to produce articles such as hollow doll parts, syringe bulbs, and hollow flexible toys. Some parts molded from plastisols are shown in Fig. 13-8.

Rotational molding of plastisols involves the same process as described under rotational molding of powdered resins, page

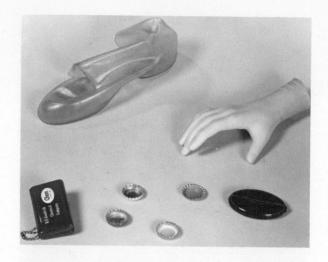

Fig. 13-8. Typical plastisol products made by using slush and dip molding processes.   (The B.F. Goodrich Co.)

143. The principal differences are the mold is charged with a liquid rather than a powder, and the resulting products are flexible. Typical plastisol rotational molding products are basketballs, footballs, automotive arm rests, toys, and industrial parts.

PLASTISOL MOLDING IN THE LABORATORY

Using plastisols in the plastics laboratory provides an insight into many processes

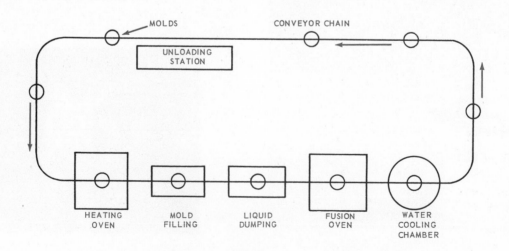

Fig. 13-7. Schematic drawing of a slush molding, conveyor type, production line.

used in industry. Clear and colored plastisols are readily available for molding and coating purposes. A heating oven, with temperature controls, is generally the only equipment required for most processes, shown in Fig. 13-9.

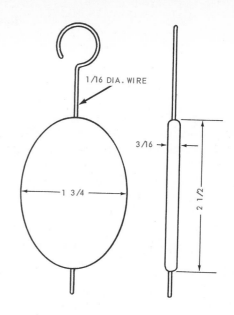

Fig. 13-10. Working drawing of a mold for a coin purse made from 3/16 in. aluminum sheet.

Fig. 13-9. Laboratory heating oven with temperature controls, air circulating fan, and heat range to 550 deg. F. (Quincy Oven Co.)

DIP MOLDING IN THE LABORATORY

Many product design possibilities are adaptable to the dip molding process. Eyeglass cases which are flexible and furniture tips are typical examples. Fig. 13-10 shows a drawing for a plastisol dip molded coin purse.

The molds for dipping are usually made from sheet or cast aluminum. Aluminum is easy to form and retains heat well. The mold should have some attachment, such as a wire, to be held while dipping and fusing. If the inside of the product requires fine detail, the internal mold should be well polished. The sequence for dip molding a coin purse is as follows:

1. Preheat mold in an oven at 350 deg. F. for 10 min.

2. Remove mold from the oven and quickly immerse in an open container of plastisol. Allow to dwell for 1 1/2 min.

3. Slowly withdraw the mold from the plastisol at an even rate to obtain a smooth surface, Fig. 13-11. Allow excess plastisol to drip into the container.

4. Return the mold to the oven at 350 deg. F. for 8 min. for final fusing, Fig. 13-12.

Fig. 13-11. After withdrawing the heated mold plastisol, it is allowed to drip to remove the

5. After curing, remove the mold from the oven and quench in cold water.

6. With a knife and straightedge, carefully slit the plastisol coating down the center from one end to the other, Fig. 13-13.

7. Open the plastic at the slit and peel the coating from the mold as shown in Fig. 13-14.

8. With a sharp knife, trim the ends of the coin purse where the plastisol built up on the pins.

The completed coin purse, along with other designs, is shown in Fig. 13-15.

In cases where it is necessary for the plastisol coating to stick to the mold, Fig.

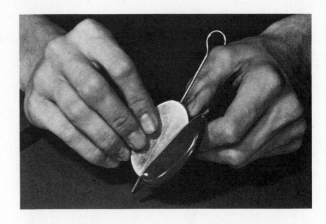

Fig. 13-14. Stripping the molded purse from the aluminum mold.

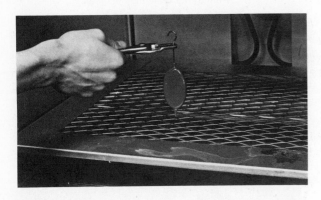

Fig. 13-12. Hanging the mold in the oven for final fusion, after dipping.

Fig. 13-15. Completed coin purses with alternate designs.

Fig. 13-13. Splitting the plastisol purse with knife and straightedge.

ROTATIONAL MOLDING IN THE LABORATORY

The process of rotational molding of plastisols is similar to molding with powders, page 143. The amount of plastisol placed in the mold will determine the wall thickness of the product. Rotational molding time and temperatures are similar to those used in dip molding of plastisols. Rotational molding of plastisols is desirable when a high degree of flexibility is required in the product, Fig. 13-17.

13-16, a primer should be applied prior to preheating. This will adhere the plastisol to he product.

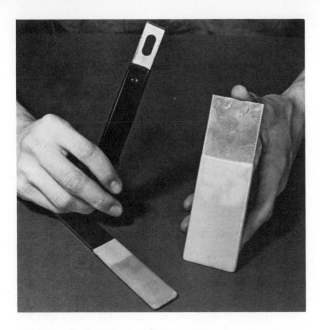

Fig. 13-16. Test molds and plastisol-coated wrench handle.

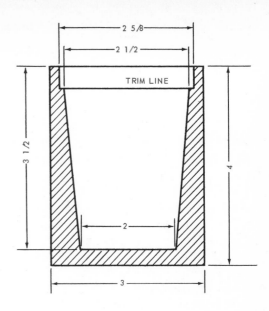

Fig. 13-18. Cross section of a mold for slush molding made from bar aluminum. A thick wall mold is a good heat retainer.

Fig. 13-17. Photo which shows flexibility of rotationally-molded plastisol ball.

with gloves. Fill mold with plastisol to the correct level, Fig. 13-19.

3. Allow the plastisol to fuse on the inside of the mold for 2 min. Empty the remaining plastisol back into the container. (If the mold loses too much heat during the dwell time, leave it in the oven.)

4. Return the mold to the oven for 5 min. at 350 deg. F. to fuse the inside of the product wall.

SLUSH MOLDING IN THE LABORATORY

Laboratory processes involving slush molding of plastisols can be accomplished with little equipment. Molds of many design shapes can be made from aluminum. See Fig. 13-18. The process is the same as used by industry except each operation is carried out by hand. The procedure for slush molding is:

1. Place the mold in the laboratory oven at 350 deg. F. for 10 min. for preheating.

2. Remove the hot mold from the oven

Fig. 13-19. Filling the heated mold with plastisol to correct level.

5. Remove the mold from the oven, quench in cold water, and strip the product from the mold.

6. Trim the top of the product with a sharp knife to the correct height and your slush molded product is ready for use.

TEST YOUR KNOWLEDGE – CHAPTER 13

1. Plastisols are a mixture of finely ground _____ in a suitable liquid plasticizer.
2. Dip molding makes use of which one of the following types of molds:
   a. An internal mold.
   b. A hollow mold.
   c. An external mold.
   d. None of the above.
3. Industrial dip molding systems usually include five stations along the conveyor line. They are:
   a._____.
   b._____.
   c._____.
   d._____.
   e._____.
4. Plastisols are used in spray coating systems for:
   a._____.
   b._____.

   c._____.
5. The term dwell means _____.
6. Rotational molding of plastisols produces a _____ product.
7. Slush molding makes use of a _____ filled with plastisol.
8. Wall thickness in slush molding is determined by _____ and _____.
9. Plastisol spray coatings are possible up to thickness of:
   a. .010 inch.
   b. .050 inch.
   c. .100 inch.
   d. None of the above.
10. Cured plastisols have the properties of being a _____.
11. Typical products of the dip molding process include:
    a._____.
    b._____.
    c._____.

SUGGESTED ACTIVITIES

1. Design and construct an aluminum mold for dip molding plastisols. Test the mold by carrying out the complete molding process.
2. Write to manufacturers of plastisols and secure literature on the types they make. Prepare a report on the various grades of plastisols produced. Include properties and recommended uses.
3. Secure some tools that should have plastisol coated handles. Determine the type of plastisol best suited for each and perform the coating operation.
4. Using a mold from your plastics laboratory and plastisol, make a slush molded product.
5. Prepare a paper on the uses of plastisol molding for automotive parts. Write to manufacturers requesting literature.
6. Construct a bulletin board display which illustrates typical products made by rotationally molding of plastisols. Write to manufacturers requesting illustrations for use in your display.

Chapter 14
LAMINATING PLASTICS

A laminated material is made by bonding together two or more layers of material to form a single unit or sheet. Plywood is a good example of such a material. Sheets of wood veneer are glued together to form a thicker, more rigid sheet. The same principle is involved in laminating plastics.

In laminating plastics, the individual layers may be sheet plastic; or, layers of other materials such as paper, fabric, and wood may be included with the plastic.

The individual layers of the laminate may be bonded by using a synthetic adhesive, or

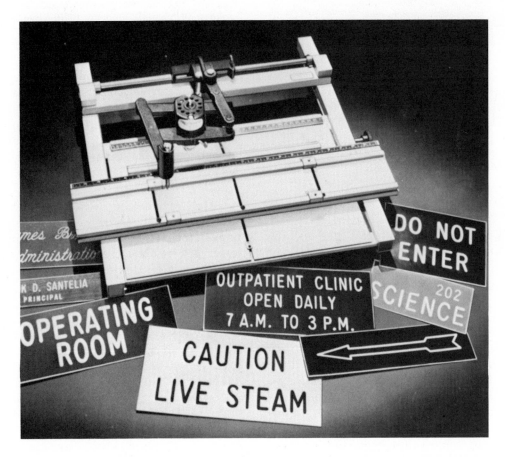

Fig. 14-1. Laminated nameplate material is made by laminating colored surface sheets over a center sheet of another color. Letters are formed by cutting through the top sheet and exposing the center layer. (Scott Machinery Development Corp.)

by fusion of the layers of plastic. See Fig. 14-1. In another type of laminate, individual layers are impregnated with a synthetic resin which bonds the layers together.

A laminate formed by pressures exceeding 1,000 psi is known as a high-pressure laminate. Formica used for kitchen cabinet tops is a good example of a high-pressure laminate. Laminates involving molding pressures of 0 to 1,000 psi are known as low-pressure laminates. Laminated credit and identification cards are examples of low-pressure laminates, Fig. 14-2.

Fig. 14-3. This multiple platen hydraulic press is used to make small laminated products. (Clifton Hydraulic Press Co.)

Fig. 14-2. Laminate made from a clear sheet of cellulose acetate bonded to a printed paperboard backing. This is an example of low-pressure laminating. (Celanese Corp.)

THE LAMINATING PROCESS

Most laminating is done in a heated multiple platen hydraulic press. See Fig. 14-3. The layers of material are usually impregnated with a thermosetting resin and allowed to dry. Typical resins used for impregnating are the phenolics, melamines, silicones, epoxies, and polyesters. The impregnated layers are assembled to the proper thickness. Then, they are placed in the laminating press between polished plates. As heat and pressure are applied, the resin flows throughout the layers of material forming a solid laminated mass. Laminating temperatures range from 300 to 350 deg. F. Pressures range from 1,000 to 2,000 psi. When thermosetting resins are used, the laminate cures in the press in a few minutes. If done carefully, it can be removed from the press while still hot.

Laminating thermoplastic sheet is done in a similar manner except that after the sheets are fused together by using heat and pressure, the platens must be cooled before removing the product. Longer cycle periods are required as the press must be reheated before the next laminate can be prepared.

High production decorative plastic laminates are made with melamine impregnated paper top sheets and phenolic impregnated kraft paper cores. Rolls of kraft paper are saturated with liquid phenolic resin monomer as they are pulled through a resin bath. The paper is then dried and cut to size. The same operation is done with the melamine and a light weight decorative (printed) paper. All of the required sheets are then stacked together. A protective transparent melamine sheet is placed on top of the decorative layer, followed by numerous kraft paper layers, Fig. 14-4. The stack is covered with release paper and then a polished stainless steel plate. Many stacks are loaded into the multiple platen press where they are processed at the same time. After the normal thermoset curing period, the completed lam-

Fig. 14-4. Note the various layers that make up a decorative high pressure plastic laminate. (Formica Corp.)

decorative paper layer. Paper is the normal substrate because of its toughness and low cost. Cloth fabrics, such as canvas, cotton and glass, are used in sheet laminates in combination with phenolics and epoxies. These high strength composites are used for electrical insulating parts, fuse boxes, terminal boards and printed circuit boards, Fig. 14-4B. Some laminated sheet is also machined or punched to form cams and gears. Decorative laminates are commonly used for furniture, wall and ceiling panels. A back-lighted ceiling is shown in Fig. 14-5.

inate is removed hot. It is then trimmed to final sheet size.

LAMINATED PRODUCTS

Flat sheets are the most commonly laminated products. They are widely used for counter tops and cabinetwork, Fig. 14-4A. Many colors and patterns are printed on the

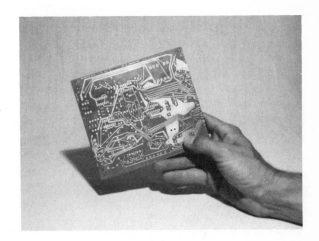

Fig. 14-4B. This circuit board is laminated from canvas sheets impregnated with epoxy resin.

Fig. 14-4A. These custom designed kitchen cabinets and counters are done in black and white plastic laminates. (Formica Corp.)

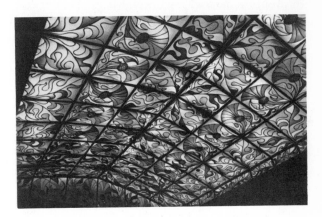

Fig. 14-5. Laminated ceiling panels made from woven glass cloth impregnated with phenolic resin. (Hooker Chemicals & Plastics Co.)

LAMINATING IN THE LABORATORY

A number of laminating processes can be performed in the plastics laboratory. Pre-impregnated papers, of plain color and with wood grain, and clear overlay sheets can be laminated using heat and pressure. These materials are the same as used in industry to produce high-pressure laminates. They are usually impregnated with phenolic or melamine resins and require only heat and pressure in a hydraulic press to complete the curing. Instructions supplied by manufacturers should be carefully followed.

Laminating plastic sheet material provides good laboratory experience involving a number of processes and materials. Polyvinyl chloride sheet is generally used to laminate flexible, thin materials such as identification cards, photographs, and important papers. Clear and colored vinyl sheet in thicknesses of .010 to .020 in. is generally used for this type of laminating. Acrylic sheet is often used where the laminate requires greater strength and rigidity, or when thicker specimens such as coins or jewelry are to be enclosed in the laminate.

A standard hydraulic press with heated platens and a water cooling system may be used quite satisfactorily. A laminating press designed for this specific purpose will provide a quick cycle. See Fig. 14-6. The laminating process consists of preparing the laminate sandwich, which includes the necessary plastic and paper sheets, heating the sandwich under pressure to fuse the sheets together, and cooling the laminate. A typical lay-up of a laminating sandwich is shown in Fig. 14-7.

The step-by-step procedure for making a plastic laminate is as follows:

1. Select the polyvinyl sheet and insert to be laminated. Cut to size, and assemble as shown in Fig. 14-8.
2. Set temperature control dial at 310 deg. F. When the heat indicator light goes out, the press is ready for the laminating cycle.
3. Insert the laminating sandwich between

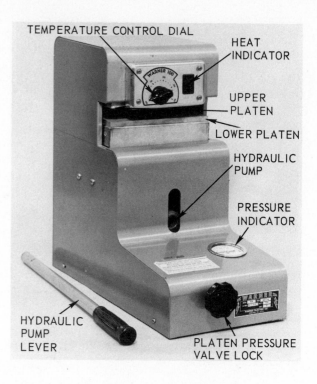

Fig. 14-6. A laboratory laminating press with heated platens and water cooling system.

press platens and lightly tighten the pressure valve lock, Fig. 14-9.

4. Insert the pump arm and pump the arm up and down until the platens make a light pressure on the laminate. The sandwich should preheat at this slight pressure for one minute.

5. After the preheat period, increase the platen pressure to 300 psi for two minutes,

Fig. 14-7. Cross section of the layers of a polyvinyl chloride laminating sandwich.

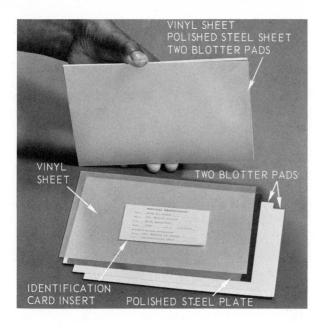

Fig. 14-8. Assembly of the layers of a laminating sandwich with a photograph insert.

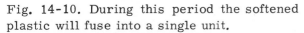

Fig. 14-9. Inserting the laminate sandwich between the press platens. Note temperature setting of 310 deg. F.

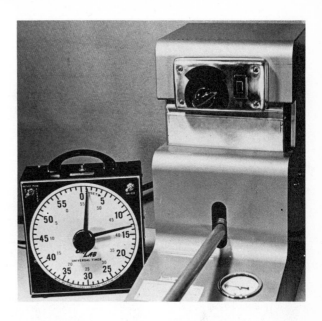

Fig. 14-10. Timing the fusion phase of the laminating cycle.

Fig. 14-11. Removing laminate sandwich from the press.

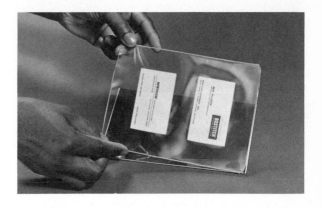

Fig. 14-12. Separating plastic sheet from the polished backing plate.

Fig. 14-10. During this period the softened plastic will fuse into a single unit.

6. At the end of two minutes, turn the temperature control dial to 150 deg. F. and turn on the water cooling system.

7. When the heat indicator light goes on again (at the 150 deg. F. setting) the pressure should be released on the platens and the laminate removed, Fig. 14-11. Turn off the cooling water.

8. The laminate should be carefully separated from the polished plates, Fig. 14-12.

9. Trim the laminate to the desired size as shown in Fig. 14-13.

10. The corners may be rounded, Fig. 14-14. The completed laminate is shown in Fig. 14-15.

Fig. 14-15. Completed laminates with rounded corners.

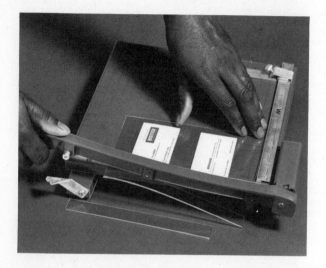

Fig. 14-13. Trimming laminate on a paper cutter.

to 15 min. depending on sheet thickness and the number of sheets used in the laminate.

COMPREG LAMINATING

Compreg is a laminate made from layers of wood veneer impregnated with a thermosetting resin and cured under heat and pressure. The name is derived from compression (com) and impregnate (preg).

The laminating process consists of soaking 6 x 6 in. (or other size) sheets of veneer (1/28 in. thick) in a solution of phenol formaldehyde resin until saturated. This requires about 24 hours. The veneers are re-

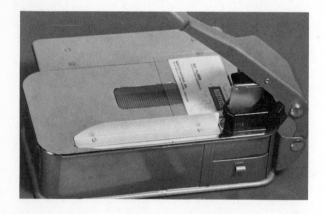

Fig. 14-14. Rounding corners of laminate on a corner trimmer.

The same steps are followed in laminating acrylic sheet except the temperature setting should be 320 deg. F. at a pressure of 200 psi. Pressure periods will vary from 5

Fig. 14-16. A stack of 24 sheets of phenolic impregnated veneer ready for laminating and a sample of compreg.

Fig. 14-17. Compreg knife handles provide the beauty of wood with the durability and hardness of phenolic resin.

processing. The desired number of sheets for the compreg laminate are placed in the press with a platen temperature of 300 deg. F. A pressure of 800 psi is applied to the laminate for 12 min. The press may then be opened, without cooling, and the compreg laminate removed. Cooling is not necessary since the phenolic resin is thermosetting and becomes a hard, rigid material when polymerized.

moved from the liquid and allowed to dry for another 24 hours at which time they are ready for laminating, Fig. 14-16. A standard laboratory hydraulic press can be used for

During the laminating process the phenolic resin liquefies, due to the heat, and flows throughout the layers of veneer before solidifying. This seals the layers into one solid, dense mass. Compreg is used commercially for items such as airplane propellers, forming dies for sheet metalwork, knife handles, buttons and other products. See Fig. 14-17.

TEST YOUR KNOWLEDGE - CHAPTER 14

1. Most laminating is done on a _____ .
2. The term compreg is taken from the two words _____ and _____ .
3. Examples of low-pressure laminating are _____ and _____ .
4. The principle of high-pressure laminating is to produce a _____ sheet material.
5. Five materials commonly used as the layers for a laminated structure are:
   a. _____ .
   b. _____ .
   c. _____ .
   d. _____ .

   e. _____ .
6. Common thermosetting resins used to impregnate laminate sheets are:
   a. _____ .
   b. _____ .
   c. _____ .
   d. _____ .
   e. _____ .
7. Forming pressures of over _____ psi are usually used to form high-pressure laminates.
8. _____ sheet is used to laminate important papers and photographs.

SUGGESTED ACTIVITIES

1. Collect samples of laminated plastics. By examination, try to determine what material was used for the laminate reinforcement and what resin was used.
2. Select paper or other thin specimen and laminate it between sheets of polyvinyl chloride in the laboratory press.
3. Using embossed sheet metal, make a set of two plates to provide a textured surface for a plastic laminate.

4. Prepare a written report describing resins and types of laminated plastics used in the electricity-electronics field.
5. Obtain sheets of wood veneer, impregnate them with phenolic resin and make a sheet of compreg.
6. Write to a company that manufactures high-pressure laminates and request pamphlets on their products for your laboratory technical library.

Constant, sequential checks are performed during sheet molding compound (SMC) production to secure high quality parts.   (The Budd Co.)

Chapter 15
REINFORCED MOLDING

In using the reinforced plastic molding process a product is formed by using a plastic resin and a woven reinforcing material. The reinforcing material is usually saturated with resin prior to molding. Many different techniques are used.

Reinforced plastics molding is commonly known as fiber glass molding. However, the term fiber glass molding correctly refers only to products involving a resin and some form of glass fiber reinforcement.

DANGER! The catalyst or hardener commonly used with a polyester resin for fiber glass work is very dangerous. If one drop of catalyst enters your eye, eye tissue will be progressively destroyed. Unless the catalyst is washed from the eye within four seconds, there is no known way of stopping destruction of the eye. Blindness can result! Always wear eye protection when working with a polyester resin catalyst.

Typical products of reinforced molding processes are automobile bodies, ceiling and wall panels, pole vaulting poles, aircraft sections, boats, and safety helmets, Fig. 15-1.

Two primary molding methods are closed mold processing and open molding. Closed mold processing requires either matched molds, Fig. 15-2, or half molds in which some other form of pressure is used, such as a pressure bag or vacuum bag. The use of closed molds is very similar to straight compression molding. However, in this process, the reinforced resin goes into the mold

Fig. 15-1. Production of reinforced plastic power boats is conveyorized at the final inspection stage. (Molded Fiber Glass Companies)

as a liquid with the reinforcing material. This is generally known as fiber reinforced plastics, FRP. The other major method is to pre-mix the plastics resin with the reinforcing materials as a bulk molding compound (BMC), sheet molding compound (SMC), or thick molding compound (TMC).

203

Fig. 15-2. These are two halves of a matched mold for a glass fiber reinforced automotive part.
(Efficient Industries Corp.)

Open molding refers to hand lay-up or spray lay-up in which a half mold is covered with a reinforcing material and soaked with resin, allowed to cure, and removed from the mold.

RESINS FOR REINFORCED PLASTIC MOLDING

Thermosetting resins are primarily used in the reinforced molding process. Polyesters are the leading resins used since they provide good strength, are comparatively low in cost, and cure at room temperatures. These are used extensively for building panels, boats, and automobile parts. Diallyl phthalate, silicone, phenolic, epoxy, and melamine resins are also used, taking advantage of the special properties of each.

REINFORCEMENTS FOR MOLDING

Reinforcing materials are of a variety of types. Glass fibers are the leading materials, since they can be used in so many ways. These are produced as woven mats of varying thicknesses and types of weave.

They are also available as random mats with glass fibers running in all directions. This type mat is easier to stretch around intricate parts of a mold cavity. Chopped glass fibers are used in premix molding in which the resin is mixed with fibers to produce a paste. These are also mixed with a resin and sprayed on the mold surface.

Other reinforcing agents include asbestos mats and paper which provide high heat resistance, strength, and insulation properties. Cloth and plastic fabrics, paper, and graphite are used for special applications.

REINFORCED PLASTICS MOLDS

Various materials are used for molds for reinforced plastics. High production matched molds are usually made of steel, Fig. 15-3. These molds produce a finished surface on both faces of the product. Open molds are made of materials such as concrete, sheet metal, plaster, plastics, and wood. These molds produce parts with only one finished surface. The type of mold used is determined by the production method, quantity of products required, and the plastic resin to be used. These molds are mounted on the upper and lower platens of appropriate size hydraulic presses, Fig. 15-4.

Fig. 15-3. Preparation of a matched mold by tracer milling for a mat reinforced truck housing. The pattern for the mold is at the top.

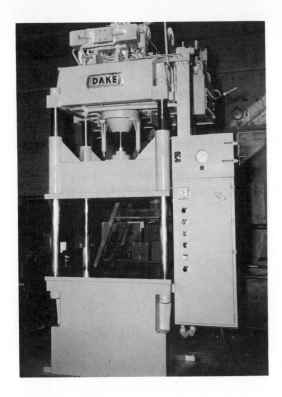

Fig. 15-4. This 75 ton capacity heated platen hydraulic press with timer and controls is used for fiber reinforced plastics (FRP) molding.   (Dake Corp.)

MATCHED MOLDING PROCESS

Matched molding, Fig. 15-5, is used for high production of plastic reinforced products. The setup shown is used for the production of glass fiber reinforced polyester chair seats, Fig. 15-6.

Preforms for matched molding are prepared from chopped glass fibers by air felting. Air felting consists of spraying the chopped fibers against a rotating screen shaped to the contour of the mold, Fig. 15-7. The thickness of the preform is determined by the amount of glass fibers sprayed. The preforms are removed from the forms and are ready for molding.

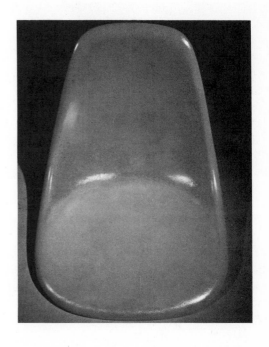

Fig. 15-6. Glass fiber reinforced chair seat.

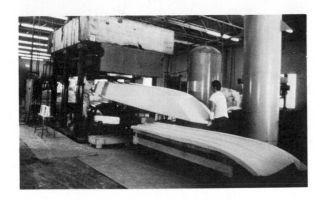

Fig. 15-5. Note matched molds on a hydraulic press for the production of fiber reinforced boat hulls.
(Molded Fiber Glass Companies)

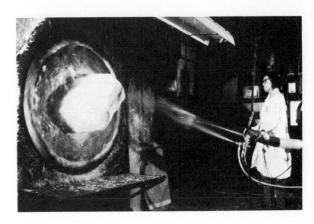

Fig. 15-7. During air felting, chopped fibers are sprayed on rotating screen.

The reinforced molding process begins with a quality control inspection of the preform to assure uniformity of the mat. Polyester resin, with the catalyst added, is weighed and poured onto the preform as it rests on the lower half of the mold, Fig. 15-8. The resin is spread by hand with a spatula to cover most of the surface. Using

about eight minutes, Fig. 15-10. Fig. 15-11 shows the mold open and the cured seat being lifted from the lower half of the mold, using a section cup. Fig. 15-12 shows removing of the flash. The sharp edges of the seat are then sanded smooth. Polishing the finished product is shown in Fig. 15-13. Quality control on a production scale of this type insures that all parts with imperfections will be rejected.

Fig. 15-8. Weighed resin is poured onto the preform.

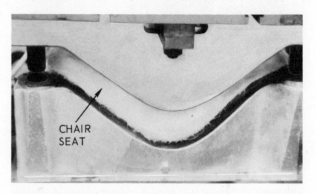

Fig. 15-10. The top half of the mold closes on the preform and is heated to 245 deg. F.

a highly polished two-part matched mold is important to the smoothness of the finished product and also to the ease of removal of the part from the mold, Fig. 15-9.

Buttons are pressed to close the mold, and the chair seat cures at 245 deg. F. for

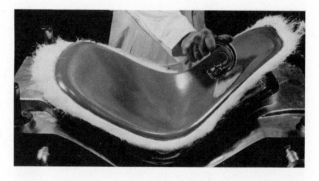

Fig. 15-11. A suction cup is used to lift the fully cured seat from the mold. Asbestos gloves may also be used to protect your hands from the 245 deg. F. heat.

SHEET MOLDING COMPOUND (SMC)

Sheet molding compound is a mixture of liquid polyester resin, fillers and randomly distributed glass fibers produced in sheet,

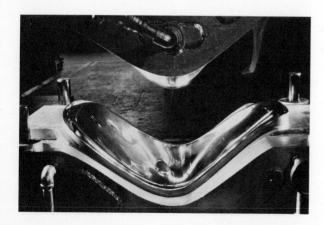

Fig. 15-9. Highly polished reinforced plastic steel mold. (Cincinnati Milacron Co.)

use up to 12 tons of glass per day in the preparation of SMC. Molding quality of the material is determined by a spiral flow test that is properly done under heat and pressure, Fig. 15-15.

SMC molding is performed on specially designed hydraulic presses, fitted with matched molds. The SMC material is cut to

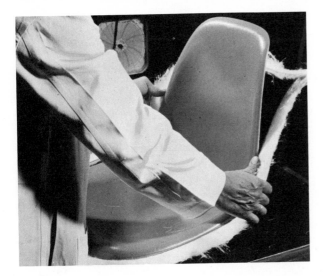

Fig. 15-12. Flashing is stripped away from the molded part by the operator.

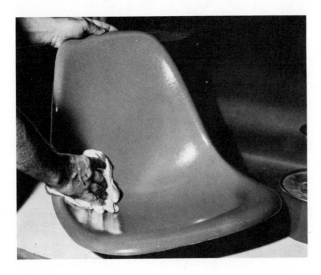

Fig. 15-13. Final polishing and inspection for imperfections in the surface. (Cincinnati Milacron Co.)

Fig. 15-14. The sheet is then rolled onto a mandrel and allowed to mature (partially polymerize) for 24 hours. It is then ready for molding. A typical formula for SMC is a suitable polyester resin with 200 parts per hundred calcium carbonate filler combined with 28 percent 2 inch (51 mm) glass fibers, randomly distributed. Some companies may

Fig. 15-14. Top. This continuous 48 inch (122 cm) wide SMC machine produces up to 2500 pounds (1125 kg) of material an hour. Bottom. The glass fibers, cut in one inch (2.6 cm) lengths, are automatically sandwiched between layers of resin paste. (Molded Fiber Glass Companies)

Fig. 15-15. Compression molded spiral flow test piece measures the moldability of SMC. (The Budd Co.)

shape and thickness of the part. Fig. 15-16 shows the part being removed from the press.

SMC processing is now used in many large applications, such as computer consoles, office furniture, seating, tractor and truck parts and sports vehicles. High production of SMC parts has become a major feature of the automotive industry, Fig. 15-17. The major advantages of SMC over other materials has been higher output, enhanced appearance, improved performance and lower cost. Other SMC materials are formulated from polyurethane based resins and epoxies for specific applications.

size and measured for exact weight. The material is quite soft and flexible at this stage. It is then positioned on the metal mold and cured under heat and pressure. Pressures vary from 100 to 2000 psi (690 to 13790 kPa), temperatures from 225 to 330 deg. F. (107 to 166 deg. C) . Curing cycles run from 1/2 to about 5 minutes depending on the size,

Fig. 15-17. Large automotive parts are rapidly produced on this 3000 ton SMC molding press. (The Budd Co.)

Fig. 15-16. Cured SMC part is removed from the upper half of mold. (Molded Fiber Glass Companies)

BMC materials are quite similar. They are prepared by blending very short glass fibers, up to 1/2 inch (12.7 mm) long, into a polyester putty to which fillers have been added. BMC is available in bulk form or as extruded preforms cut to weight for molding. The high filler content of BMC materials makes it possible to produce rigid parts with good electrical properties and at a low cost. Typical applications include distributor caps, motor brush holders and other products requiring good heat resistance.

Thick molding compound, TMC, is much the same as SMC except that the glass fiber distribution can be improved in a random manner. This is due to the method of production and thickness of the material. TMC is produced up to 2 inches (51 mm) thick compared to the 1/4 inch (6.4 mm) thickness of SMC. Many molding problems can be eliminated by using the thicker material, especially flowability in complex shapes.

HAND LAY-UP PROCESS

The open mold, hand lay-up process is the simplest and oldest type of reinforced molding, Fig. 15-18. It is ideal for use in prototype or low volume production of large parts such as furniture, boats and swimming pools. A number of hand tools are required for processing, as shown in Fig. 15-19. The first step is to carefully prepare the mold and then apply the pigmented gel coat of polyester resin that completely covers the mold surface. See Fig. 15-20. The mold is usually made of plaster, metal, wood or reinforced plastics. When gel coat has thickened, a mat of woven glass fibers is manually placed on the mold. It is then sprayed with a coat of polyester resin and smoothed out with a

Fig. 15-19. Major tools used in hand lay-up include polyester sprayer, power sander, rollers, brushes, knives and thickness gages. (Owens-Corning Fiberglas Corp.)

roller or squeegee to force out air bubbles, Fig. 15-21. Layers of glass fibers reinforcement and resin are added to build up the desired thickness. Either catalysts or accelerators can be added to the resin to allow the resin to cure without external heating. Final finishing of the open side of the product is

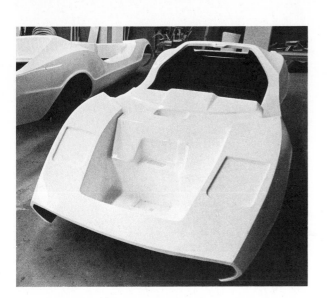

Fig. 15-18. Strength and style of complex shapes can be achieved through glass fiber reinforced molding like this automobile body. (Owens-Corning Fiberglas Corp.)

Fig. 15-20. Gel coat of polyester resin is being applied by spray gun. (Owens-Corning Fiberglas Corp.)

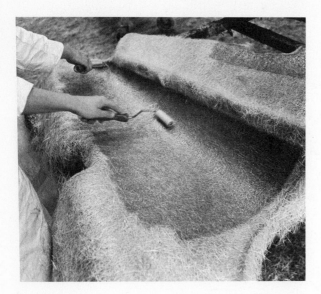

Fig. 15-21. Roll out removes any entrapped air and smooths the surface. (Owens-Corning Fiberglas Corp.)

done by sanding and the application of another resin gel coat. When the product is removed from the mold, it can be assembled with other parts by bonding or standard fasteners, Fig. 15-22.

SPRAY LAY-UP PROCESS

Glass fiber reinforced spray lay-up is similar to the hand process. The main difference is in the method of spray-up. Spray-up involves feeding continuous rovings of glass fibers through a chopper. The chopper cuts the fibers to a predetermined length and then propels them into a resin stream. Look at Fig. 15-23. The resin and glass are deposited simultaneously on the mold surface. The chopper-spray gun equipment is shown in Fig. 15-24.

Fig. 15-23. Top. The two pot spray gun mixes resin that has been promoted with catalyzed resin. Fiber chopper is at the top. Bottom. Catalyst injection spray gun uses promoted resin and adds catalyst during spraying. The chopped fibers are mixed from the top. (Binks Manufacturing Co.)

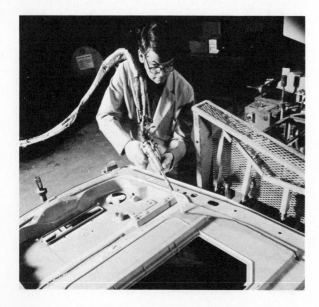

Fig. 15-22. Parts assembly can be accomplished by a tough, long-lasting, adhesive bonding process. (The Budd Co.)

are added to build the required thickness. Spray-up is particularly well suited to efficient fabrication of complex shapes. Gel coating will minimize the visibility of fiber patterns. The part is allowed to cure at room temperature. Smaller parts can be partically cured in an oven to speed up the process. When the part has become sufficiently rigid to retain the original shape, it is removed from the mold, Fig. 15-26. Some manufacturers trim the product with a knife while it is still wet. Others use power grinders and routers to trim excess material, Fig. 15-27.

Fig. 15-24. Catalyst, promotor, resin and glass fiber rovings are all fed to the spray gun on the right. (Binks Manufacturing Co.)

The mold is prepared as in the hand lay-up with mold release. Polyester gel coat is applied. Woven glass fiber roving can be added for specific strength, Fig. 15-25, but is not used in many spray lay-up products. Additional layers of chopped glass fibers and resin

Fig. 15-26. Large glass fiber reinforced part is removed from the mold with a hydraulic lift. (Owens-Corning Fiberglas Corp.)

Fig. 15-25. Woven glass fiber roving strengthens the reinforcement. (Owens-Corning Fiberglas Corp.)

The advantages of the spray lay-up process are simple, low cost tooling, portable equipment for on-site fabrication, low cost molds, accommodation of complex forms, reverse curves and large parts.

Processes which involve the same type of lay-ups but require pressure for a smooth surface on both sides of the product are the pressure bag system, Fig. 15-28, and vacuum bag process, Fig. 15-29. These also aid in forcing out trapped air bubbles and provide a uniform thickness to the lay-up.

Fig. 15-27. Excess material is being trimmed with a portable router using saw blade. (Owens-Corning Fiberglas Corp.)

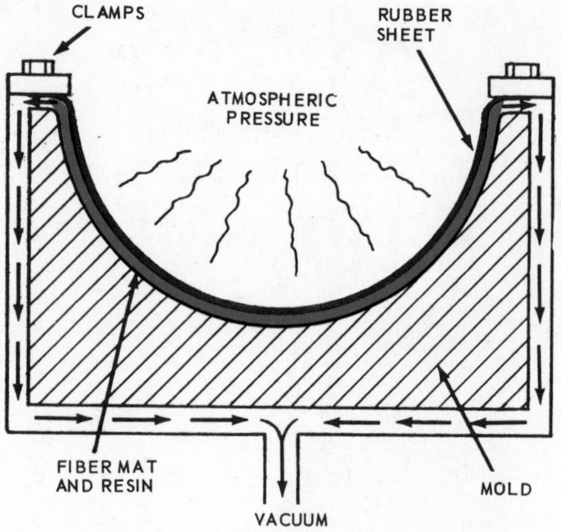

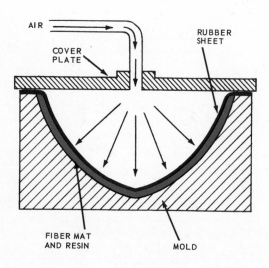

Fig. 15-28. Reinforced fiber molding using a rubber pressure bag.

Fig. 15-29. Vacuum bag process for molding a fiber reinforced lay-up.

REINFORCED MOLDING IN THE LABORATORY

A large number of reinforced plastics processes can be performed in the laboratory following the same techniques used in industry. Popular resins are the polyesters and epoxies since they are available in liquid form and are ready to use when mixed with a catalyst. Instructions for mixing and handling provided by the manufacturer should be followed. Danger! Always follow safety rules when using catalyst for polyester resin. Reinforcing materials commonly used are glass fiber random and woven mat, decorative cloth fabrics, and woven material such as flannel.

Molds for reinforced plastics are easily made of wood, plastics, plaster, and aluminum. The mold material often depends on the process used and the quantity of products desired from the mold.

MATCHED MOLDING IN THE LABORATORY

Molds for the matched molding process may be purchased or constructed in the laboratory. Fig. 15-30 shows a matched mold for a large salad bowl. The procedure for making the reinforced molding is as follows:

1. A heavy coating of paste mold release is applied to each half of the mold. This is allowed to dry for a few minutes, and is lightly polished with a cloth.

2. Two circular discs of glass fiber random mat and one disc of a decorative fabric are prepared.

3. A gel coat of polyester resin is prepared according to manufacturer's specifi-

Fig. 15-30. This matched mold for a 16 in. diameter salad bowl was cast from aluminum.

cations and is applied to each half of the mold with a brush.

4. While the gel coat is becoming tacky, a batch of polyester resin is mixed for the lay-up.

5. One disc of glass fiber mat is placed in the lower half of the mold and saturated with resin, Fig. 15-31.

6. When this is completed, the decorative cloth is placed in the mold followed by the second disc of fiber mat.

7. The total lay-up is then saturated with

resin, the top half of the mold set in place, and a heavy weight placed on the mold.

8. After a curing period of about three hours, the mold halves are separated and the product removed, Fig. 15-32.

9. The molding should be allowed to cure overnight before bandsawing the flash.

10. The sharp edges are smoothed with a file or sandpaper and the molding is completed, Fig. 15-33.

HAND LAY-UP MOLDING IN THE LABORATORY

A hand lay-up or stretched fabric reinforced product can be produced in a number

Fig. 15-32. Separating the mold halves with a wood pry stick.

Fig. 15-31. Coating glass fiber mat with polyester resin.

Fig. 15-33. Filing the edge of the completed molding.

of ways. One method is to construct a wood mold using dowel rods or other wood supports over which the fabric is draped. Such a mold is illustrated in Fig. 15-34. After

Fig. 15-34. A wood mold for a stretched fabric reinforced lay-up. Mold release is applied to the contact surfaces.

the mold is constructed, masking tape is placed over surfaces where the lay-up will make direct contact and a mold release is applied over the tape. A plain or decorative fabric is then stretched over the mold and stapled in place, Fig. 15-35. If a white fabric is used, the resin may be colored to the desired shade.

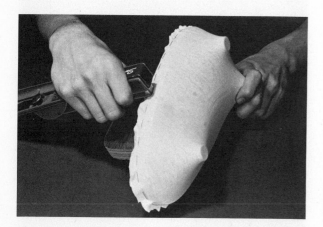

Fig. 15-35. Stapling the reinforcing fabric to the edges of the mold.

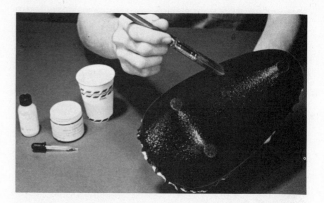

Fig. 15-36. Applying a coat of polyester resin to the reinforcing fabric.

Either polyester or epoxy resin may be used for the plastic coating. An estimated amount of resin is poured into a container and catalyst added according to the manufacturer's specifications. Coloring is added at the same time and the compound is stirred until completely mixed. A light coat of resin is then applied to the fabric as shown in Fig. 15-36, and allowed to cure

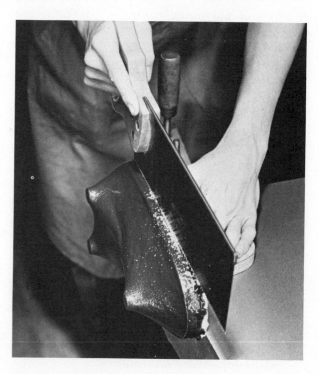

Fig. 15-37. Using a backsaw to cut the plastic reinforced candy dish from the mold.

Reinforced Molding

Fig. 15-38. A completed fabric reinforced candy dish.

overnight. Two more similar resin applications are made to provide for a good wall thickness. When the last coat has cured, the reinforced plastic product is cut from the mold, Fig. 15-37. A final coat of resin may be applied to the inside of the dish to provide a smoother surface if desired. The sawed edges are sanded smooth to complete the candy dish, Fig. 15-38.

Woven glass fiber mat, as well as many other fabrics, may be used for reinforcement. Solid wood or plaster molds may be used in the same manner, being sure the entire mold surface is well coated with mold release.

TEST YOUR KNOWLEDGE - CHAPTER 15

1. The two primary reinforced molding processes are _____ and _____.
2. The term _____ is commonly used in place of reinforced plastics molding.
3. Open molds for reinforced plastics are often made of:
   a. Plaster.
   b. Wood.
   c. Plastics.
   d. Sheet metal.
   e. All of the above.
4. Matched molding is used for _____ of reinforced plastics products.
5. Two methods of preparing preforms from chopped glass fibers for matched molding are:
   a. _____.
   b. _____.
6. Resins most commonly used for hand lay-ups with glass fiber mat are:
   a. _____.
   b. _____.
7. The principle of using a pressure bag for some reinforced molding is to _____.
8. The two types of spray lay-up systems involve:
   a. _____.
   b. _____.
9. Regardless of the reinforcing process being used, a _____ should always be applied to the mold before the operation begins.
10. Typical products made by reinforced molding are:
    a. _____.
    b. _____.
    c. _____.
    d. _____.
    e. _____.
11. Why are polyesters the leading resins for reinforced molding?
12. A _____ or _____ is commonly utilized to squeeze out bubbles and to smooth out the resin coating in a hand lay-up process.
13. Sheet molding compound (SMC) is formulated from:
    a. _____.
    b. _____.
    c. _____.
14. The two types of chopper-spray guns used in reinforced molding are:
    a. _____.
    b. _____.

SUGGESTED ACTIVITIES

1. From the technical literature in your laboratory, prepare a paper describing the types of reinforcing materials used with each major plastic resin for industrial reinforced molding.

2. Design and construct a matched mold to be used for a fiber reinforced product to be constructed in your laboratory.

3. Write to a manufacturer of fiber glass reinforced boats for literature on their products. Ask for samples of the reinforcing materials they use and ask if they have pictures of the reinforced molding process they use. Make a display around the materials and pictures you receive.

4. Devise a simplified chart to indicate the amount of catalyst to be used with various amounts of polyester reinforcing resin. Base your chart on resin volume from one ounce to one quart.

5. Secure samples of different products made by reinforced plastics molding. Prepare a display panel showing the objects and how they are used.

6. Secure information on automatic mixing and spraying equipment for applying polyester and epoxy resins. Write a brief paper explaining the use of this equipment.

7. Design and construct a one piece mold for a reinforced hand lay-up. Mix the appropriate resin and make a lay-up to test the mold.

8. Write a report on the use of reinforced plastics in the automobile industry. Contact manufacturers and obtain technical literature for reference.

9. Plan and give a demonstration on the different ways reinforcing resin may be colored.

10. Prepare a safety poster illustrating the precautions to be observed when mixing a catalyst with a resin.

Operator is removing foamed high density polyethylene product from molding area of a 200 ton vertical structural foam molding press.   (Williams-White and Co.)

Chapter 16
PLASTIC FOAM
MOLDING

Foamed plastics (also called expanded or cellular plastics) are made by adding air or gas to a plastic resin to form a sponge-like material. Modern developments in plastic foam molding have placed this among the major processing techniques of the industry. See Figs. 16-1 and 16-2. Resins commonly used in the production of plastic foam include polystyrene, polyurethane, polyethylene, cellulose acetate, epoxy, silicone, and phenolic.

Plastic foams may be classified as to material, type, cell structure and density.

The two principal types of plastic foam are rigid and flexible: rigid are resistant to crushing; flexible are easily crushed.

Fig. 16-2. This kitchen cabinet door is made from rigid polyurethane foam. It resembles oak but will not warp or expand and contract.

Fig. 16-1. Expanded foam shipping cases for delicate instruments. The packaging industry makes use of the outstanding properties of many plastic foams.
(Koppers Co.)

Cell structure refers to the openings in the foam. Plastic foams may have either an open or closed cell structure. Open cell structure indicates the foam has interconnected cells or openings running through the material. Foam with an open cell structure will allow a liquid to pass through like a sponge. This kind of foam structure is usually flexible. Closed cellular structure means the cells within the plastic foam are separate air or gas filled units. The closed structure will not allow liquid to pass through the foam. Closed structure foams are usually rigid.

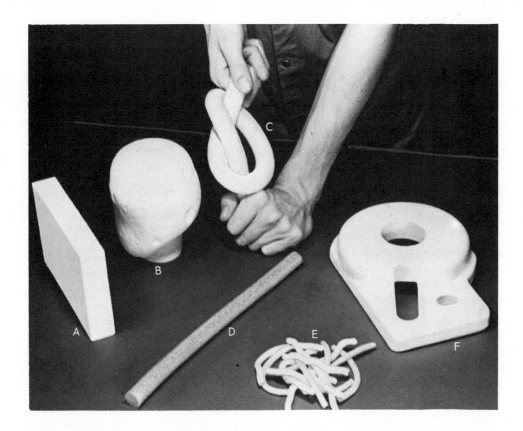

Fig. 16-3. A—Extruded polystyrene foam slab. B—Rigid polyurethane foam. C—Flexible polyethylene foam rod. D—Rigid cellulose acetate foamed rod. E—Expanded polystyrene foamed spaghetti. F—Molded polystyrene foam.

The density of a plastic foam is indicated by its weight in pounds per cubic foot. Density of foams ranges from 0.10 to 70.0 lbs./cu. ft. The density of a foam is important for product uses in terms of weight, insulating properties, and flotation. A number of plastic foams are shown in Fig. 16-3.

METHODS OF FOAMING

Plastic foams are produced by three principal methods . . . mechanical, physical, and chemical.

In MECHANICAL FOAMING of a plastic resin the action is similar to preparing a milk shake. The resin, as a solution or emulsion in liquid form, is vigorously agitated until it becomes a foam of air bubbles. Fusing of the foamed resin by heat causes it to remain a solid foam. Polyvinyl chloride

plastisols can be mechanically foamed in this manner.

PHYSICAL FOAMING usually involves the use of compressed gasses or chemicals which change their physical form during the foaming process. Foamed polyethylene, for example, is produced by forcing compressed nitrogen gas into molten polyethylene while the resin is under pressure. As the pressure on the molten resin is released, either in a mold or an extruder, the nitrogen expands and foams the plastic.

CHEMICAL FOAMING, in simple terms, involves mixing or dissolving a chemical compound in a liquid resin which will react, usually under heat, to form a gas. The gas released within the molten or liquid resin causes the foaming action. Foamed polyurethanes are produced by this method.

CHARACTERISTICS AND APPLICATIONS OF FOAM

Foamed plastics may be classified into five major areas of use:

1. Flotation.
2. Packaging.
3. Cushioning.
4. Insulation.
5. Structural.

The flotation properties of some plastic foams make them ideally suited for many purposes. Because of the closed cell structure, these foams will not absorb water. Typical uses include water skiing life belts, buoys, boat and airplane pontoons. Another advantage is they cannot be punctured like inflated rubber or plastic flotation devices.

Foams have a major use in the packaging industry. Since they can be produced in a variety of shapes and forms which cushion against shock, and are light in weight, they make ideal packaging materials. See Fig. 16-4.

Fig. 16-4. Tomatoes are cushioned from damage during shipping in this formed plastic crate. (Packaging Systems Corp.)

Plastic foams have replaced many other materials in "cushioning" applications. Flexible foams find uses in furniture seats, aircraft and automobile crash padding, arm rests, and carpet padding. Polyurethane and polyethylene foams are especially suited to cushioning applications, Fig. 16-5.

Fig. 16-5. Flexible urethane foam is molded to shape for use in these automotive seats. (Union Carbide Corp.)

The insulating properties of foamed plastics are outstanding, both for hot and cold conditions. This property is typically illustrated in the use of polystyrene foam coffee cups. Heat from hot coffee can barely be felt through foamed plastic which is less than 1/8 in. thick. The closed cell structure also makes the cups waterproof. Foam insulation applications include refrigerator and freezer linings, construction wall panels, picnic coolers and food chests, and refrigerated railroad cars.

Structural foams have a dense skin and cellular core. The strength to weight ratio of these materials is high enough to be classified as structural. Commonly found in 1/4 to 1/2 inch (6.4 to 12.7 mm) thicknesses, structural foamed parts range in application from machine housings to giant shipping pallets.

Characteristics of the various types of foamed plastics are listed in the reference tables.

FOAM MOLDING PROCESSES

Industrial foam molding processes are numerous. In general, foams are produced in slabs, boards, and sheets for further fabrication. Some foams are molded and extruded while others are foamed in place.

EXTRUDED FOAM

Extruded foamed plastics account for the bulk of boards, sheets, and rods produced. The process involves the same principles as regular plastic extrusion except a blowing agent or gas is introduced into the extruder to foam the hot resin before it leaves the die. Most extruded foam plastics are stock materials and require further fabricating into a final product. Extruded foam boards, for example, may be cut, shaped, and cemented to form almost any desired shape.

By far, the most common extruded foam product is flat sheet made from polystyrene. Flat sheet is best extruded from a tubular die. As it emerges from the die, it expands uniformly, unlike foam extruded from a flat sheet die. The process is performed with two extruders. In the first extruder, the plas-is heated. Then, the blowing agent is injected in through the side of the cylinder, Fig. 16-6. If it were extruded out of the die at this point, the material would be hot enough to cause cell rupturing. The blowing agent would expand too fast. It is therefore pumped into a second extruder to cool the melt. From here, it travels through the die, expands and is pulled over a sizing mandrel. It is usually slit in two, pulled through a series of rollers and wound up. Most foamed sheet is vacuum formed into packaging containers such as egg cartons, Fig. 16-7, meat packaging trays and fast food containers.

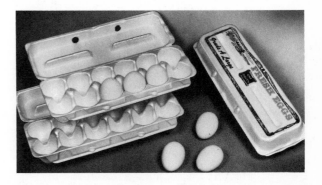

Fig. 16-7. Foamed polystyrene extruded sheet has been vacuum formed to produce these shock resistant, self-locking egg cartons. (Plastic Carton Corp.)

STRUCTURAL FOAM MOLDING

The structural foam molding process is similar to injection molding of solid thermoplastic resins. Foaming is achieved by introducing inert gas (usually nitrogen) directly into the melt or by using chemical blowing agents premixed with the resin. See Fig. 16-8 for a schematic diagram of a foam sheet extrusion line. In this case, heating plastic in the cylinder releases the gas which disperses into the melt. When the mixture is shot into the mold, the gas expands, forming a cellular structure with a tough solid skin, Fig. 16-9. Structural foam molded parts typically have a swirl surface pattern.

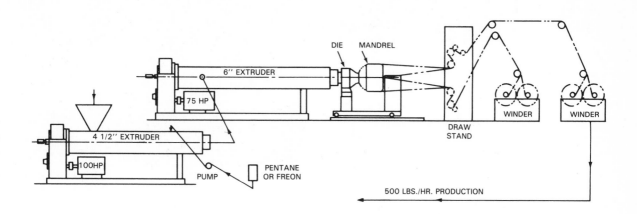

Fig. 16-6. Schematic diagram of foam sheet extrusion line. This combination of extruder sizes has been found to work well together. (NRM Corp.)

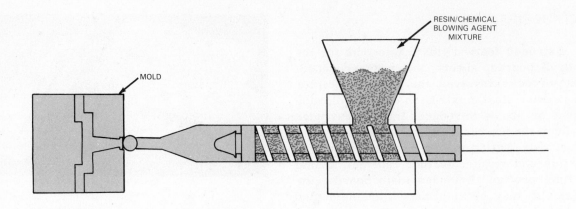

Fig. 16-8. This is structural foam molding on a reciprocating screw injection machine. (General Electric Co.)

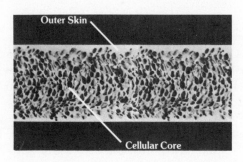

Fig. 16-9. Cross sectional view of a structurally foamed product wall section showing structure of skin and core. (General Electric Co.)

Large parts are made on special machines equipped with accumulators, Fig. 16-10. An extruder feeds the molten resin containing the blowing agent into the accumulator (or pressure chamber) where it is held under pressure to prevent expansion. When the predetermined amount of resin has been stored, the valve is opened and a piston forces the melt into the mold cavity, Fig. 16-11. Parts over 100 lbs. (45 Kg) are possible with this process. Structural foam machines, like the one in Fig. 6-12, can handle the clamping pressures for these parts. Pressure is not exceedingly high. This is due to low injection pressures and resultant cavity pressures of around 500 psi (3448 kPa).

Many different plastics are used in this process. Volume plastics such as polystyrene or polyethylene are used for garbage containers, picture frames, furniture parts and other wood grain products. Engineering resins like polycarbonate, ABS, PPO and thermoplastic polyesters are specified for applications requiring strength and good surface

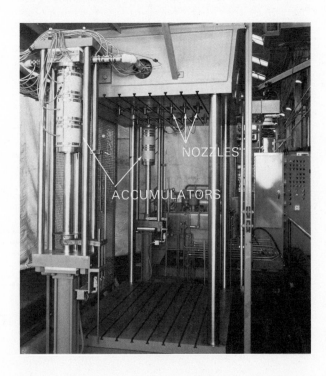

Fig. 16-10. This large structural foam molding machine has two accumulators and multiple nozzles for rapid mold filling. (Williams-White and Co.)

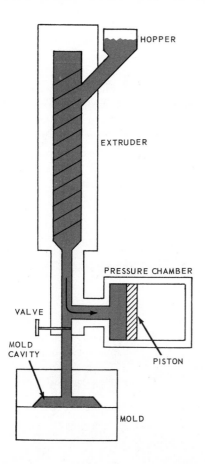

Fig. 16-11. Schematic of a structural foam molding press. The foamed shot has been made by the piston, the valve closed, and resin for the next shot is being extruded into the pressure chamber.

hardness. Common products include molded chairs, car body parts and housing for machinery and office equipment.

Structural foam molding offers several advantages over standard injection or reinforced molding. Large parts are practical with moderate machine cycle times. The cellular structure provides good strength to weight properties along with material savings. Previously designed multicomponent metal objects can be molded as one part with this process. Drawbacks include large machine investments. Also, parts may need painting to cover the swirl pattern found on the surface.

CASTING FOAMS

Liquid resins containing catalysts and chemical blowing agents are used in casting and foaming-in-place processes. The foams are generally produced by mixing liquid resin with catalyst and a chemical blowing agent. As the chemical reaction takes place, the mixture expands to form a cellular structure. Polyurethane foams, both flexible and rigid, and polyvinyl chloride plastisols, are primarily used in these processes.

The foam casting process is similar to the casting of any liquid resin. A two-part

Fig. 16-12. Typical foam molding machine with high volume mold capacity and low clamping pressure requirements. This press is rated at 300 tons.   (Dake Corp.)

aluminum mold is generally used with an opening at the top to allow for expansion. (Open mold casting is seldom done, since no pressure is developed on the foamed resin and it expands in an uncontrolled manner.) In this process, the two halves of the mold are separated and coated with a mold release. The two liquid components of the polyurethane resin are mixed to the correct proportions and poured into the lower half of the mold. The mold is quickly closed, clamped, and the foaming resin expands to fill the mold cavity, Fig. 16-13. Typical cast foamed plastic products are duck decoys, life saving rings and floats, and numerous toys.

Foaming-in-place is similar to casting except that the mold is an actual product, such as a refrigerator door, and the resin is foamed within the cavity of the product.

Polyurethane is the major foam used in this process in both flexible and rigid forms. The commercial process makes use of a polyurethane mixing and dispensing unit as shown in Fig. 16-14. The dispensing machine is used to mix the resin components and force the foamed resin, under pressure, into the interior cavity of the product. Once within

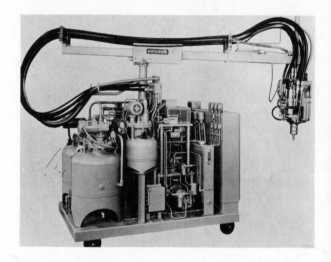

Fig. 16-14. This polyurethane foam dispensing unit permits fast filling of foaming-in-place cavities. (Borg-Warner Chemicals)

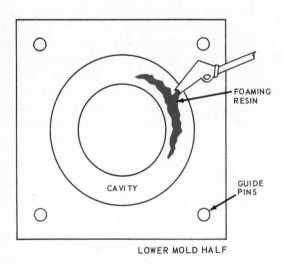

LOWER MOLD HALF

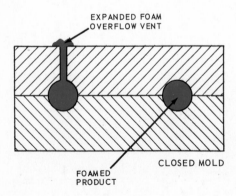

Fig. 16-13. Schematic of foamed polyurethane life ring casting. Filling the mold cavity with foaming resin is shown at the top, while a cross section of the expanded life ring is shown below.

the cavity, the foamed resin expands and solidifies. Large parts are removed by hand, Fig. 16-15, and flash is trimmed. Fig. 16-16 shows a modern foaming-in-place application. Other uses include insulating the walls of refrigerated railway cars, automobile dash pads and arm rests, and flotation compartments in boats. The process eliminates the need for molds and makes it possible to get insulation and cushioning into intricate cavities of a product.

EXPANDABLE FOAM MOLDING

Polystyrene is the principal resin involved in the manufacture of expandable beads. Gas

Fig. 16-15. A two component liquid urethane resin was used to make this automobile seat cushion. (Union Carbide Co.)

ing cases, Fig. 16-17; and the smaller beads for articles like drinking cups. Molds are usually made of aluminum or stainless steel. On a production basis, they are mounted in presses which open and close for loading and fusion cycles.

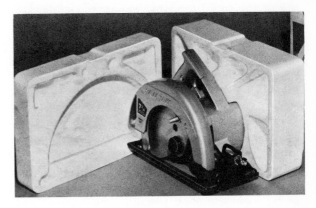

Fig. 16-17. Expanded polystyrene packaging case made from large expandable beads. (Koppers Co.)

is contained in each tiny, hollow pellet or bead. When heat is applied, the gas or blowing agent causes the bead to expand. At the same time, the styrene skin softens, allowing the bead to blow up like a small balloon. Since the skin is soft and molten as the bead expands, it will fuse to other beads. If enough beads are placed in a confined mold and heated, they will expand to fill the mold cavity and provide enough pressure to "weld" the bead together to form a rigid foam. This is a closed cell structure and, after molding, is called an expanded foam.

Expandable polystyrene beads are available in various graded sizes. The larger sizes are used for molded blocks and packag-

A typical molding sequence includes pre-expansion of the beads, filling the mold with pre-expanded beads, subjecting the mold to heat, cooling, mold opening, and ejection of the product.

Pre-expansion of the beads is done by using steam, radiant heaters, hot water, or oven heat. This brings the beads to almost the density of the final product, Fig. 16-18.

Fig. 16-16. A two-part injection molded ABS chair seat. Flexible, liquid polyurethane foam is forced into the cavity and expanded. Other parts of the chair are wood.

Fig. 16-18. These pre-expanded beads are ready for molding. (Arco Polymers, Inc.)

Commercial pre-expanders can be adjusted to provide the required density up to 20 lbs./cu.ft.

From the pre-expander, the beads are fed into the mold, either by gravity filling or by compressed air blowing. The molds are completely filled with pre-expanded beads. Heating of the mold for final expansion and fusion of the beads is done by forcing steam through holes in the mold, placing the mold in an autoclave (large pressure cooker), or pumping high-pressure steam into a tightly sealed mold.

Upon final expansion, the mold is cooled by circulating cold water through cooling tubes or spraying water on the mold. The expanded foam product is ejected from the mold by a blast of air pressure, knockout plugs, or stripper plates. The majority of products molded from expandable polystyrene beads are used for insulating and packaging purposes.

Expandable polystyrene in lace and spaghetti forms is used extensively for packaging. See Fig. 16-19. These materials are expanded by radiant heat or steam to provide excellent lightweight, shock absorbent packaging materials.

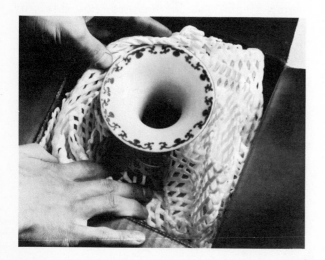

Fig. 16-19. Expanded polystyrene foam lace used for packaging. (American Hoechst Corp.)

FOAM MOLDING IN THE LABORATORY

Two molding processes involving foamed plastics which are practical for laboratory activity are molding expandable beads, and casting liquid foaming resins. Each process is performed in the same manner as commercial molding except for the variations in equipment.

EXPANDABLE BEAD MOLDING

T requirements for expandable bead molding include the construction of the mold, usually aluminum, and a source of heat or steam. Molds may be made from cast or machined aluminum which machines easily and conducts heat well, Fig. 16-20. The

Fig. 16-20. A cast aluminum mold for an expandable polystyrene fish float.

heating requirements may be met by placing the mold in boiling water, forcing steam through a mold from a tube attached to a steam generator, or placing the mold in a steam chamber (autoclave). The following procedure is suitable for molding expandable polystyrene beads in the laboratory:

1. Expandable polystyrene beads are pre-expanded by placing them in boiling water, using a radiant heater, or a heat gun with a pre-expansion unit, Fig. 16-21.

2. The mold is filled with pre-expanded

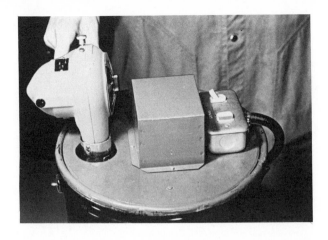

Fig. 16-21. Pre-expanding polystyrene beads using a heat gun. A rotary mixer keeps the beads in the chamber separated.

Fig. 16-22. Filling coffee cup mold with pre-expanded polystyrene beads.

Fig. 16-23. Inserting plug in the clamped mold.

beads, Fig. 16-22, being sure the complete mold cavity is full.

3. The mold halves are clamped together tightly, and the plug is inserted in the filler hold, Fig. 16-23.

4. The charged mold is carefully placed into a container of boiling water for an expansion and fusing period of 10 to 15 minutes, Fig. 16-24. (Gloves should be used to handle the mold and safety precautions observed in handling boiling water.)

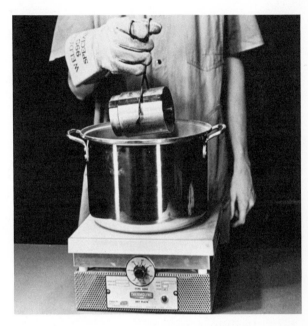

Fig. 16-24. Lowering mold into a tank of boiling water.

5. After expansion, the mold is removed from the boiling water, cooled with a water spray, and the product is removed as in Fig. 16-25.

An aluminum coffee cup mold and some finished products are shown in Fig. 16-26.

Casting of either rigid or flexible polyurethane foam is done in the same manner as described for the commercial process. Specifications for mixing the two component polyurethane resins supplied by the manufacturer should be followed. Metal and plaster molds are suitable for use in casting.

Fig. 16-25. Removing molded foam cup from the male half of the mold.

Fig. 16-26. Completed polystyrene foam coffee cups.

TEST YOUR KNOWLEDGE - CHAPTER 16

1. Plastic foams have either an _____ or a _____ cell structure.
2. Foamed plastics are also called _____ or _____ plastics.
3. Plastic foams are classified as to:
   a. _____ .
   b. _____ .
   c. _____ .
   d. _____ .
4. The type of foam refers to its properties of being either _____ or _____ .
5. Which of these plastics are used for foaming?
   a. Polyurethane.
   b. Polyethylene.
   c. Cellulose acetate.
   d. All of the above.
   e. None of the above.
6. The density of a plastic foam is indicated by _____ .
7. Foam molding processes can be classified as:
   a. _____ .
   b. _____ .
   c. _____ .
   d. _____ .
8. List the five classifications of foamed plastics as to use:
   a. _____ .

   b. _____ .
   c. _____ .
   d. _____ .
   e. _____ .
9. Structural foam molding is similar to _____ of solid resins.
10. Most structural foam molding machines are equipped with large _____ for holding the molds and _____ for injecting a large amount of plastic at one time.
11. Foamed sheet material extruded for packaging applications is usually made of _____ .
12. The three methods of producing foams are:
    a. _____ .
    b. _____ .
    c. _____ .
13. Describe "foaming-in-place."
14. Commercially, _____ is used to cause the final expansion of expandable polystyrene beads in the mold.
15. What causes expandable polystyrene beads to expand?
16. Two foams primarily used for casting processes are:
    a. _____ .
    b. _____ .

SUGGESTED ACTIVITIES

1. Design, construct the mold, and pour a rigid or flexible polyurethane foam casting.
2. Prepare a series of safety posters on the correct way to handle molds and equipment during molding of expandable polystyrene beads.
3. Prepare a list of technical terms dealing with foamed plastics and their processing. Give a clear and concise definition of each.
4. Develop a research problem to investigate the foaming agents used with different plastic resins. Request literature from manufacturers and use your technical library to obtain the necessary information, then make a chart listing the resins, the foaming agents used, and the method in which the foam expands.
5. Using a mold from your plastics laboratory, pre-expand the correct volume of expandable polystyrene beads and mold a product.
6. Write to manufacturers for literature on the extrusion of foamed plastic sheet. Make a schematic drawing which illustrates the complete process. Label each piece of equipment and describe each step in the process.
7. Set up an experiment to determine the density (lbs. per cubic foot) of a given foamed plastic material.
8. Prepare a display of products made from foamed plastics. Indicate the plastic foam used and method of processing for each product.

Typewriter ribbon cassettes are ultrasonically welding on automated rotary table machine.
(Ultra Sonic Seal Co.)

Chapter 17

FABRICATING

PLASTICS MATERIALS

A large segment of the plastics industry deals with the fabrication of standard shapes (rods, sheets, tubes, etc.) and of molded products. Fabrication generally refers to the cutting, shaping and fastening of plastic parts made from stock material, Fig. 17-1. It also refers to the assembly of molded plastic products which require various types of fastening devices. See Fig. 17-2.

This chapter will deal with cementing, welding, machining and other processes used to fabricate plastic products. Since commercial and laboratory fabricating techniques are similar, the processes are grouped together.

CEMENTING PLASTICS

Many fabricating problems involved with the fastening together of plastics parts can be solved by using adhesives.

Bonding plastic materials together falls into two groups, cohesion and adhesion. Cohesion involves the softening of the surfaces to be joined by using a liquid which will dissolve or soften the resin. As the surfaces are softened and the plastic parts pressed together, the molecules of both parts intermingle. When the liquid solvent evaporates, the joint acts like a single piece of plastic as the molecular structure returns to its original form. Solvent cementing does not add a new material to the joint. It might be compared to fusion welding of steel, in which the two parts are melted and allowed to mix together before cooling.

Liquids used in solvent cementing are usually solvents which will dissolve the plastic surface, or monomers of the same resin being used. Monomer cements dissolve the surfaces, and also polymerize (react chemically) with the plastics to be bonded to form the joint. A typical application of solvent cementing is shown in Fig. 17-3. Some solvent cements have a small amount of the plastic to be cemented dissolved in them. This makes the liquid thicker so it will tend to fill variations (low spots) in the joint surface. These are referred to as bodied cements or dope cements. Some thermoplastics which have a highly crystalline structure are not easily dissolved by a solvent. Polyethylene is a good example. Amorphous plastics (plastics without crystals), such as polystyrene, polycarbonate, cellulosics, and acrylic are easily cemented by solvents. Thermosetting plastics are difficult to dissolve and do not cement well by this process.

Plastics materials joined by adhesion make use of a film of a different material which sticks to each surface of the joint. This is similar to the gluing of wood. The materials do not dissolve; instead, the adhesive sticks to both surfaces. Adhesion is due to the molecular attraction between the glue film and the materials being joined. Contact cement and silicone resins are examples of adhesives which bond by adhesion.

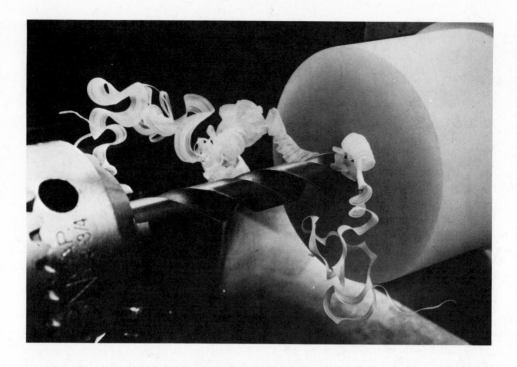

Tough nylon rod drills easily on an automatic lathe. Note continuous curled chip being formed.
(Cadillac Plastic and Chemical Co.)

Fig. 17-1. Stock sheet, rod and other standard plastic shapes stored at a plastics warehouse.
These materials are primarily used in the fabrications of products. (Westlake Plastics Co.)

Fig. 17-2. Molded products, such as this chair side, require special fastening devices to assemble the final product. (M.S. Plastics Co.)

Fig. 17-4. Capillary action draws the solvent cement across the joined surfaces when applied with a brush. A light pressure holds the joint secure.

THE CEMENTING PROCESS

Solvent type cements may be applied to the prepared plastic joint by dipping, using a small brush, Fig. 17-4, or with an eyedropper. Cement should be applied to both surfaces to be joined before bringing them together; or at the joint where capillary action will draw the liquid along the surfaces. Fig. 17-5 shows how sheet plastic may be dipped and clamped to form a strong joint.

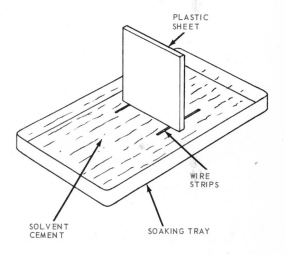

Fig. 17-3. Applying a solvent cement by brush to the two halves of a molded part.   (Borg-Warner Chemicals)

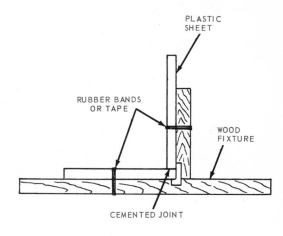

Fig. 17-5. Top. Soaking plastic sheet in solvent cement. Wires keep softened sheet from sticking to tray. Bottom. A simple fixture to hold plastic sheet while the joint cures.

Fig. 17-6. Making a square cut on PVC pipe using a miter box. (The B.F. Goodrich Co.)

Fig. 17-8. Using a rag to clean the surface to be joined.

to remove dirt, grease, and foreign matter. Cement is then applied to the outside of the pipe, Fig. 17-9, and the inside of the fitting.

A typical solvent cementing job is to join and seal PVC plastic pipe and fittings. The pipe should be cut squarely as shown in Fig. 17-6. Hand saws, power saws, and special pipe cutters may all be used successfully. After the pipe is cut, all burrs should be removed with a knife or a scraper, Fig. 17-7. Burrs left on the pipe will often cause voids (gaps), resulting in a poor joint. Joining surfaces should be cleaned with a rag, Fig. 17-8,

Fig. 17-9. Applying cement to the pipe surface that extends into the fitting. (The B.F. Goodrich Co.)

The coating should be applied rapidly and generously to all contacting surfaces. Assembly is made quickly, while the cement is still wet, by pushing the pipe all the way to the bottom of the fitting, Fig. 17-10. When the pipe is seated, a slight twist will spread the adhesive and provide a sealing action. A similar sequence may be used for other solvent cementing jobs with variations in applying cement and clamping as necessary.

Bonding thermosetting or dissimilar plastics requires the use of a different adhesive (not a solvent). Epoxy adhesives are used extensively for this purpose when a rigid joint is required. Contact cement and sili-

Fig. 17-7. Removing burrs with a knife. Both inside and outside edges should be scraped.

Fig. 17-10. Inserting the pipe into the fitting.

cone adhesives work well where some flexibility in the joint is desirable.

Hot melt adhesives are adaptable to production bonding as they are quick to cool and require no curing time. See Fig. 17-11. They are primarily used for dissimilar plastics or plastics to metal bonds.

Some cements commonly used with plastics are listed in the rear reference section of this book. Plastics manufacturers and suppliers can provide specific information on adhesives best suited for use with plastics they supply.

SAFETY PRECAUTIONS SHOULD ALWAYS BE TAKEN WHEN USING SOLVENT CEMENTS. BREATHING VAPORS OF SOME SOLVENTS MAY CAUSE TOXIC EFFECTS. WORKING AREAS SHOULD BE WELL VENTILATED. CARE SHOULD BE TAKEN TO KEEP CEMENTS FROM CONTACTING YOUR SKIN.

WELDING PLASTICS

Welding is one of the most advantageous ways of fastening plastic materials, Fig. 17-12. It is limited, however, to thermoplastic resins. In welding operations, heat must be applied to melt the resin at each surface of the joint. The source or method of providing heat classifies welding into four major groups:
1. Fusion welding.
2. Friction welding.
3. Hot gas welding.
4. Ultrasonic welding.

There is no all-purpose welding method. Each process has its own unique characteristics. Selection of a process depends on factors such as the resin to be welded,

Fig. 17-11. This speaker is fastened to the plastic grillwork by a 360 deg. F. (127 °C) bead of hot melt plastic adhesive. (Eastman Chemical Products, Inc.)

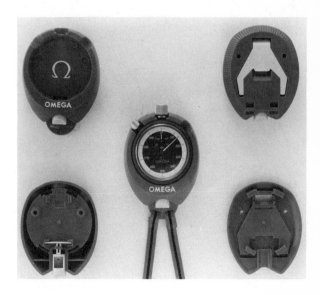

Fig. 17-12. Ultrasonic welding case of a stopwatch. The front and back of each half of the molded acetal case is also shown before welding. (Du Pont Co.)

shape of the welded part, type of weld required, quantity, and cost and availability of equipment.

FUSION WELDING

Fusion welding is a mechanical process involving the use of a hot plate or heated tool to soften the plastic to the molten state. Once the surfaces of the parts are molten, they are quickly joined together with a slight twisting motion, and held under pressure until cool. Commercially, fusion welding is used for small parts and short runs. It is particularly adaptable to welding polyvinyl chloride, polyethylene, and polypropylene. Polyethylene pipe and fittings are often welded by the fusion process.

A heated tool may be used for light welding, Fig. 17-13, or for tacking and repairs. Of greater use, is hot plate welding. The two parts to be welded are placed on the hot plate surface to be melted, Fig. 17-14. As soon as the surfaces to be joined are molten, they are quickly pressed together to fuse the two parts and form the weld, Fig. 17-15. The parts should be held together until cool. The completed joint is shown in Fig. 17-16. If a number of the same parts are to be welded, a fixture to hold them in place and apply pressure provides a more even weld.

The surface of a hot plate should be brushed with a liquid tetrafluoroethylene coating to prevent sticking of the plastic.

Fig. 17-14. Two sections of polyvinyl chloride pipe are brought to a molten state on a hot plate.

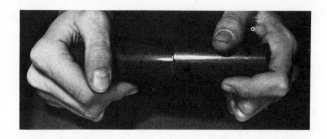

Fig. 17-15. Applying pressure to the molten surfaces of the pipe to form the weld.

Such coatings are used for repairing and coating kitchen utensils so they will not stick to foods, and are available from hardware stores.

FRICTION WELDING

Another method of developing heat at the joint of two surfaces to be welded is by rubbing them together. This is called friction welding. The friction developed causes enough heat to melt the resin and create the weld. Although it is possible to make a weld by motion other than rotational (circular), the spinning of cylindrical parts

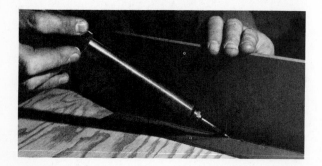

Fig. 17-13. Light welding using a heated tool to fuse the two sheets together. (Laramy Products Co.)

Fig. 17-16. Completed fusion weld on PVC pipe.

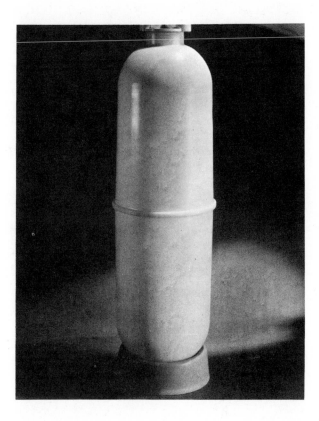

Fig. 17-17. The two halves of this pressure tank for a home water conditioner are injection molded from acetal resin and friction welded together. Note weld in center of the tank. (Celanese Plastics Co.)

against each other is the most feasible. This rotating action is also known as spin-welding. Fig. 17-17 shows a typical application of friction welding. Friction welding works satisfactorily with most thermoplastic resins. A lathe or drill press provided with a special chuck or holding fixture, may be used for spin-welding on low production jobs.

The process involves a fixture to hold the lower stationary plastic part and a spinning head to rotate the upper half of the part to be welded. Look at Fig. 17-18. As the top half of the part is spun against the lower half,

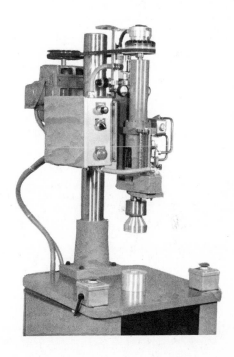

Fig. 17-18. A typical single spindle thermoplastic spin-welding machine. (HyPneuMat, Inc.)

the plastic quickly heats up, melts and forms the weld. The chuck retracts. Then, the part cools and solidifies at the weld line. Another technique uses a friction clutch which allows the moving chuck to stop when the plastic welds. Clamping pressure is held on the part until cool. Production spin-welders make use of a rotating table with a loading station and automatic ejection as shown in Fig. 17-19.

Fig. 17-19. An automated spin-welding machine welds the cap on sealed marking pens. An indexing table moves the parts to be welded and includes automatic ejection.

A laboratory spin-welder may be improvised from a standard drill press. A fixture is required to hold the stationary piece and a chuck to grip the rotating piece. Fig. 17-20 shows fixtures used to spin-weld a laboratory float. A thin coating of silicone resin is brushed on the chuck and fixture to prevent slippage. The drill press is turned on at its slowest speed and the chuck is lowered to apply pressure on the spinning

part, Fig. 17-21. As melting and fusion take place, which usually requires only a few revolutions, the chuck is released and the welded part is allowed to cool. The welded part is shown in Fig. 17-22.

Fig. 17-21. The spinning chuck rotates the two contacting plastic parts as pressure is applied.

The main advantages of spin-welding are the simplicity of equipment required and short period of time required to make welds. Most spin-welding cycles are under 30 seconds.

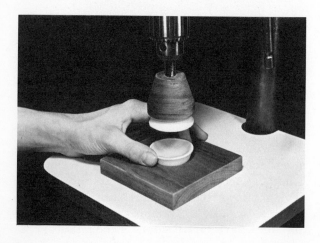

Fig. 17-20. Wood fixtures fitting the two halves of an injection molded laboratory float are mounted on a drill press.

Fig. 17-22. The welded float shows a flash at the weld line which can be trimmed if necessary.

HOT GAS WELDING

The plastics fabrication industries have seen rapid development in hot gas welding. Gas welding provides a means of constructing economically objects such as large tanks and containers, and fabricating parts in short-run production which would be difficult and expensive to mold. See Fig. 17-23. This process is also used for welding pipe joints, and in repair work.

Hot gas welding involves the use of a compressed gas. The gas is heated as it passes through a welding gun (gas does not burn), and is directed on the welding rod and pieces to be joined. A typical hot gas welding unit is shown in Fig. 17-24. When the two mating surfaces become molten, the softened filler rod fuses with them to form the weld. The schematic drawing in Fig. 17-25 shows how a hot gas welding torch operates. Beveling of the joint edges as done in oxyacetylene welding, is necessary to secure depth of weld penetration.

Resins best suited for hot gas welding are polyethylene, polypropylene, and polyvinyl chloride. However, other resins have

Fig. 17-23. A self-supporting rectangular tank made from 3/8 in. high density polyethylene. All parts were joined by hot gas welding. (Industrial Plastic Fabricators, Inc.)

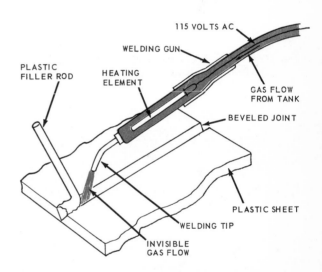

Fig. 17-25. Schematic diagram of hot gas welding. Gas flows through the tubing from the tank, around the electrical wire and heating element, and out the tip. Gas, which is invisible is shown in the drawing for purpose of illustrating the flow.

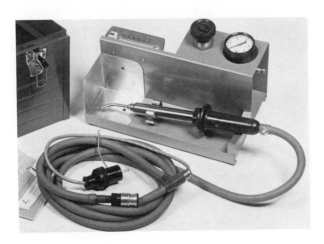

Fig. 17-24. Electrically heated welding gun. Gas passing through gun (which does not burn) is heated and directed onto pieces to be joined. (Laramy Products Co.)

been satisfactorily welded. Filler rod, usually extruded, of the same material as those pieces to be joined, should be used. Compressed air is satisfactory for welding PVC. Nitrogen is used as the heated gas on most other resins sensitive to oxygen. Temperature of the heated gas should range from 400 to 800 deg. F. as it leaves the gun. Gas pressure varies from 5 to 15 psi depending on the thickness of the material. A number of welding gun tips are available. The tips used most are the general purpose tip, tacking tip, and tape welding tip. See Fig. 17-26.

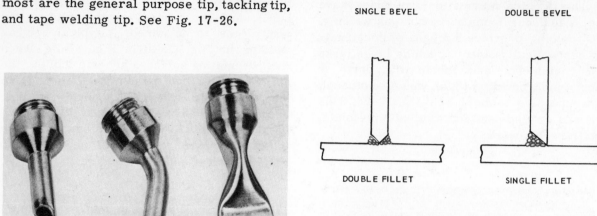

Fig. 17-27. Preparation of joint edges for flat and fillet welds.

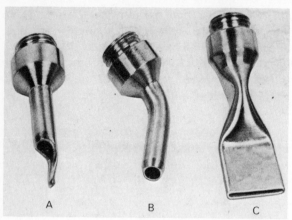

Fig. 17-26. Study hot gas welding tips. A—Tacking tip. B—General purpose tip. C—Tape welding tip. (Kamweld Products Co., Inc.)

THE WELDING PROCESS

Preparation of the pieces to be welded is an important step in the process, Fig. 17-26A. Fig. 17-27 shows the basic edge preparation

for flat and fillet welds. The procedure begins by selecting the required tip and setting the gas flow at the correct pressure, Fig. 17-28. After the gas is flowing, the heating element is turned on and the gun is allowed to warm up. Be sure to turn on the gas first, as the heating element may burn out if gas is not flowing around it at all times. In the laboratory, welding rod made on an extruder may be used, Fig. 17-29. The hot gas is directed on the joint and tip of the rod until both are molten. For polyethylene, the rod is held at an angle of about 60 deg., Fig.

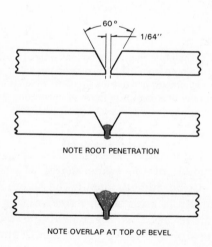

Fig. 17-26A. Notice proper edge preparation and positioning of weld beads. (Kamweld Products Co., Inc.)

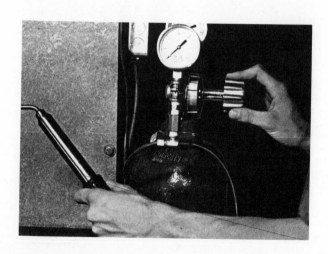

Fig. 17-28. Adjusting flow of nitrogen gas at the tank regulator.

Fig. 17-29. Polyethylene welding rod ready for use after laboratory extrusion.

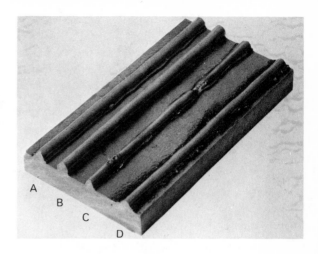

Fig. 17-31. Note condition of welds. A—Good weld. B—Lack of heat and penetration. C—Welding rod over heated. D—Too much heat applied to base material. (Kamweld Products Co., Inc.)

17-30, using a light pressure. As the rod and surfaces melt, fusion takes place and the rod deposits a bead in the joint. When welding polyvinyl chloride, the rod is held at a right angle to the surface. The weld is continued to the end of the joint at which time the rod may be pressed into the surface, heat removed, and the rod allowed to cool. To finish the surface, the rod is cut off smoothly with a knife.

Speed welding is done in a similar manner, as shown in Fig. 17-32. A speed tip is used which feeds the rod as the welding gun is drawn along the joint. The speed tip, Fig. 17-32A, has two openings, one for the rod and one for the hot gas which softens the plastic in front of the filler rod. The rod is preheated as it passes through the tip. Speed

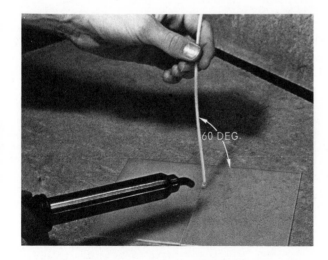

Fig. 17-30. Welding polyethylene with the rod held at an angle of about 60 deg. from the direction of the weld.

Quality welds are dependent upon the amount and location of the heat applied, Fig. 17-31. Other factors include the amount of penetration and preparation of the joint edges.

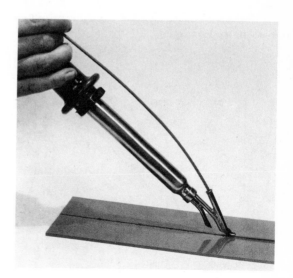

Fig. 17-32. Worker is speed welding a structural housing made of polyvinyl chloride. (Kamweld Products Co., Inc.)

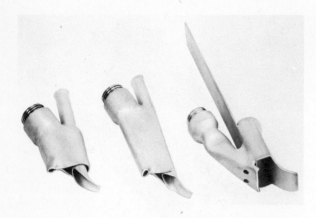

Fig. 17-32A. Three types of speed welding tips used in production hot gas welding are shown. (Kamweld Products Co., Inc.)

In all hot gas welding operations, good welds require the correct temperature, rod angle, rod pressure, and speed.

Fig. 17-33. Back welding a polyvinyl chloride socket fitting.

welding is quicker and also more accurate for production work. The proper positioning of the welding torch for a complete weld can be seen in Fig. 17-32B.

Another hot gas welding operation is to seal plastic pipe after it has been threaded or cemented. This process is called back welding and is illustrated in Fig. 17-33. This serves to seal fittings and prevent leaks. Repair operations often make use of hot gas welding, even on molded products, Fig. 17-34. Leaks and worn areas can be repaired by laying down enough beads to cover and seal the damaged spots.

ULTRASONIC WELDING

Ultrasonic welding is a high-speed production process for welding plastics by using ultrasonic vibrations. The vibrations are produced by converting electrical energy into ultrahigh frequency mechanical energy.

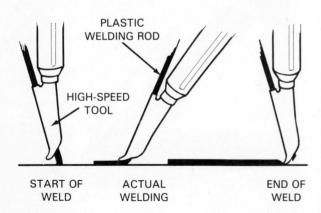

PLASTIC WELDING ROD

HIGH-SPEED TOOL

START OF WELD   ACTUAL WELDING   END OF WELD

Fig. 17-32B. These are welding positions for completing a high-speed weld. (Kamweld Products Co., Inc.)

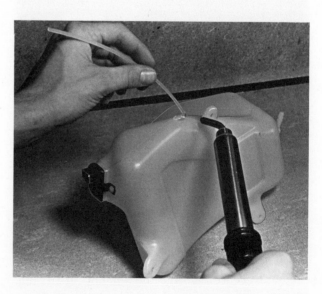

Fig. 17-34. Repairing a leak in a blow molded high density polyethylene windshield wiper cleaning fluid container.

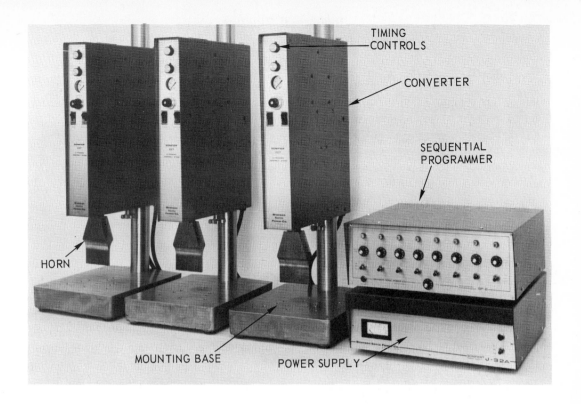

Fig. 17-35. Three ultrasonic welders with a programmer for sequential welding of parts.
(Branson Sonic Power Co.)

Such vibrations cause enough mechanical friction between the plastic surfaces to bring them to a molten state. Sixty cycles per second electrical energy from the power supply is converted to 20,000 cycles per second mechanical vibration at the horn. See Fig. 17-35. In operation, the horn is vibrating one plastic part against another which is mounted on a stationary fixture at the base. The size and shape of the plastic parts to be joined determine the size and shape of the horn. See Fig. 17-36.

THE WELDING PROCESS

Parts to be welded are mounted on a fixture attached to the base. An air cylinder brings the horn down into contact with the plastic part at a given pressure. While held in this position, ultrasonic vibrations are used to fuse the parts together. After fusion, the parts are held by the horn for a few moments during cooling, then the horn retracts. An air jet is generally used to eject the welded parts.

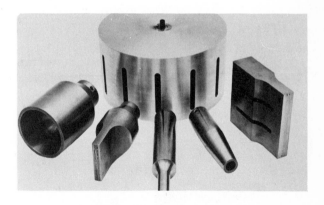

Fig. 17-36. A variety of horns used in ultrasonic welding.

There are a number of different ultrasonic welding processes, Fig. 17-36A. Standard welds incorporate energy directors molded into one half of the assembly. This is the point of contact where the plastic first starts to melt. As the pressure increases, the flow spreads completing the weld. Other processes include spot welding, staking and insertion operations.

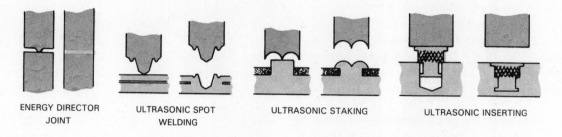

| ENERGY DIRECTOR JOINT | ULTRASONIC SPOT WELDING | ULTRASONIC STAKING | ULTRASONIC INSERTING |

Fig. 17-36A. Study the four types of welds possible with ultrasonic welding.

Flat sheets are welded together using a spot welding tip that penetrates both surfaces. Many closely spaced welds are needed for maximum strength. Staking is a riveting type process in which the head is melted and formed by the horn. Metal inserts are mated with the plastic part by ultrasonically vibrating them into an undersized hole.

Portable ultrasonic welders, Fig. 17-37, are used to assemble parts of unlimited size. They are also used to weld joints in hard-to-reach areas.

MATERIALS AND PRODUCTS

Most of the rigid thermoplastic materials are easily assembled by ultrasonic welding.

These include ABS, polystyrene, polycarbonate, ionomer, and acrylic. Softer plastics, such as polyethylene, are difficult to weld since they absorb the vibrations and do not transmit them to the secondary part of the joint. Typical products making use of ultrasonic welding are automotive lenses, dashboard components, cases for electronic components, and highway sign reflectors, Fig. 17-38.

MACHINING PLASTICS

Cutting, machining, and general shaping of plastics is required for many fabricated products and for final finishing of molded products. In fabricating, sheet stock or other

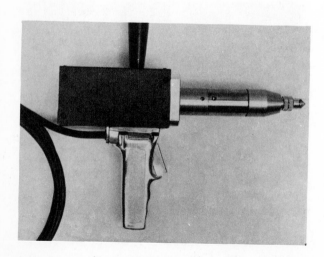

Fig. 17-37. A portable ultrasonic welder with a pistol grip handle.   (Branson Sonic Power Co.)

Fig. 17-38. These acrylic automotive lenses incorporate multiple pieces ultrasonically welded together.
(Sonics and Materials, Inc.)

standard shapes usually require cutting to size before further processing. Most of the tools and machines used for machining metals, Fig. 17-38A, and woods are also used for plastics. Variations in cutting speeds and changes in shapes or angles of cutting tools are sometimes necessary. Thermoplastics will often become soft due to the friction between the tool and material. Some thermosetting materials are so brittle they are easily chipped during machining. A brief description of general machining techniques will be given in this unit. Specific machining requirements for various plastics are listed in the textbook reference section.

Fig. 17-38A. This vertical milling machine is being used to cut out a slot. Most plastics machine well on this type of equipment.   (Cadillac Plastic and Chemical Co.)

CUTTING PLASTICS

Plastic materials are usually cut by sawing or shearing. Each process is advantageous for certain resins and certain conditions.

SAWING

Circular saw blades used for cutting wood may be used to cut plastics. However, better results can be obtained by using hollow-ground plastic cutting blades. These blades have about 7 teeth per inch, are approximately 1/16 in. thick, have a slight set, and remove about 1/10 in. of material. They usually operate at 3450 rpm. Plastic cutting blades may also be used to cut soft metals such as aluminum, brass and copper.

When sawing thermoplastic materials, remember that these materials become soft when heated. When the saw generates enough heat to melt the plastic, the saw will gum up and fail to cut satisfactorily. Reason: Plastic is a poor conductor of heat, and most of the heat generated by cutting remains at the point of cutting.

The work must be held firmly against the saw table to prevent shattering and chipping, and the saw moved slowly to prevent accumulation of a ridge of melted plastic along the edge of cut. See Fig. 17-39.

Some thermoplastic materials, particularly high-pressure laminates and filled thermosetting plastics, are so hard they require the use of carbide tipped saws.

A band saw may be used to make both curved and straight line cuts. A band saw blade dissipates the heat quicker than a cir-

Fig. 17-39. Using a radial arm saw to cut ABS sheet. (Borg-Warner Chemicals)

cular saw blade. Cutting intricate designs and curves of small radii are possible if a narrow blade is used. Skip-tooth metal-cutting blades with 4 to 6 teeth per inch are preferable, however, wood-cutting blades may be used. Be sure to move the stock into the blade slowly to avoid plastic build-up. The blade should be inspected and cleaned regularly.

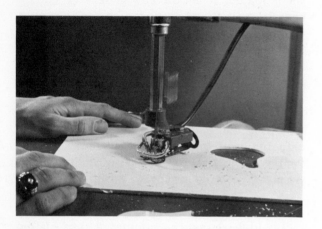

Fig. 17-40. Interior cutouts being made on plastic sheet using the jigsaw. (Borg-Warner Chemicals)

A jigsaw is useful for cutting small parts and interior cutouts, Fig. 17-40. Blades with ample set should be used. As soon as the blade fails to make a clean cut, back off and allow the stock and blade to cool.

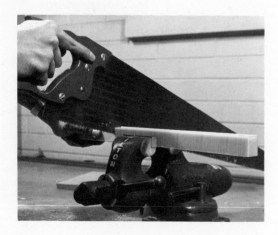

Fig. 17-41. Cutting square ABS rod with a handsaw. (Borg-Warner Chemicals)

Handsaws of almost all kinds may be used for cutoff work, Fig. 17-41. Saws used for wood are satisfactory for cutting the softer plastics. Metal-cutting hacksaws work well when cutting hard plastics. Care should be taken not to dull wood cutting saws on hard plastic materials.

SHEARING

Most thermoplastic sheet material up to 1/8 in. in thickness can be cut on a sheet metal squaring shear, Fig. 17-42. Brittle sheet, such as acrylic and polystyrene, should not be sheared as they easily crack and chip along the edges. Most thermosetting sheet or high-pressure laminates are also too brittle and abrasive for shearing. Where shearing is practical, it is often preferred over sawing because it produces smoother edges. Tin snips and paper cutters work well for small sheet work, such as preparing stock for vacuum forming.

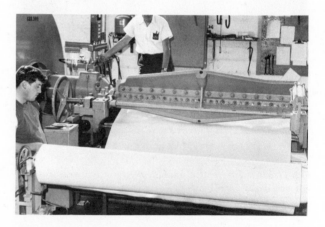

Fig. 17-42. Cutting tetrafluoroethylene sheet six feet wide on a power shear. (Cadillac Plastic and Chemical Co.)

DRILLING PLASTICS

Standard metalworking equipment is normally used for drilling plastics. The drill press, Fig. 17-43, portable power drill, and hand drill are all satisfactory. As with other machining processes, the heat build up around the drill often makes it difficult for chips to escape.

Fig. 17-43. Using drill press to drill plastics.

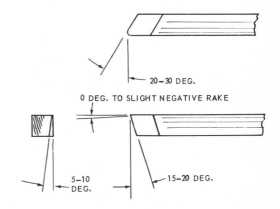

Fig. 17-45. A typical general purpose lathe turning tool for plastics.

Twist drills are available that are specially designed for plastics. Conventional twist drills are suitable if correctly sharpened. In general, drills should have maximum clearance for the cutting edges to reduce friction, and smoothly polished flutes for easier chip removal. When drilling deep holes, the drill should be backed out frequently to clear chips. Information on drill speeds and drill point angles will be found in this textbook's reference section.

TURNING PLASTICS

Plastic materials can be turned on standard metal lathes, Fig. 17-44, or automatic screw machines using the same general

Fig. 17-44. Turning threads on a bar of ABS resin in the metal lathe.

procedure used for soft metals. Heat due to friction of the turning tool against the plastic is a problem which is overcome by using light cuts and a sharp tool. The shape of a turning tool for general plastics work is shown in Fig. 17-45. The harder, tougher plastics like acetal, nylon, and tetrafluoroethylene are easier to turn than the more brittle plastics such as polystyrene and acrylic. A problem in turning many plastics are the long stringy chips which wrap around the work. Stopping the feed at intervals or pushing the chips away with a dry paint brush will usually release them from the work. Lathe tools should be mounted on center or below and have a zero to slight negative rake as illustrated in Fig. 17-45.

CUTTING FOAMED PLASTICS

Foamed plastics often require special techniques for cutting and shaping. Flexible foams are normally cut with an electric knife or a razor blade. Thin sheets may be cut on squaring shears.

Rigid foams are usually cut with conventional sawing equipment. Circular saws may be used for planks less than 3 in. thick. Production cutting of rigid foam from large billets into boards is usually done on a high-speed band saw using a skip-tooth blade. Production cutting also makes use of multiple hot wire equipment. Hot wire cutting is used on rigid thermoplastic foam, such as polystyrene, in which the wire melts the material to make the cut. A laboratory hot wire cutter is shown in Fig. 17-46. The wire, heated to a dull red by an electric current, cuts quickly both straight and in-

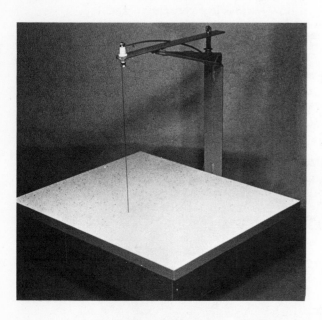

Fig. 17-46. An electrically heated hot wire cutter for expanded polystyrene foam.

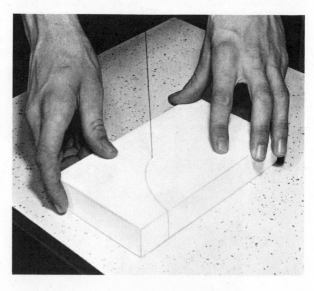

Fig. 17-47. Cutting a curved shape in foamed polystyrene using a hot wire cutter.

TAPPING AND THREADING PLASTICS

Standard taps used for metals are satisfactory if they are slightly honed to produce a rounded root on the threaded hole. Two-flute taps give more room for chip removal than the standard four-flute taps. A lubricant such as paraffin should be used and the tap frequently backed out to remove chips and keep it from binding.

tricately curved shapes, Fig. 17-47. Hot wire cutting produces a smooth surface on the foam as the melted resin tends to form a skin over the surface. Constructional details on a hot wire cutter are given in Fig. 17-48.

FINISHING AND POLISHING

Most fabricated products require finishing and smoothing to remove tool marks and scratches as a result of processing. Filing is useful in removing deep tool marks. The vixen file, a coarse, curved tooth single-cut file. Works well on thermoplastics. It is the

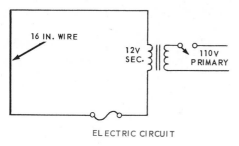

ELECTRIC CIRCUIT

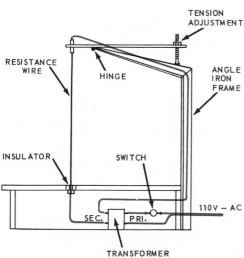

ELECTRICAL PARTS LIST

TRANSFORMER – 12V SEC. AT 6–8 AMPS
16 IN. NICHROME ® WIRE – NO. 19–20
FUSE BLOCK OR FUSE, HOLDER (OPTIONAL)
SINGLE–POLE, SINGLE–THROW SWITCH
LINE CORD AND PLUG

Fig. 17-48. Construction of a hot wire foam cutter. Table and frame may be of size desired.

type file used on soft metals in automobile body repair work. Draw filing with a single-cut mill file provides a smooth surface prior to final sanding or buffing. If sanding is necessary, a medium wet-or-dry sandpaper should be used, followed by finer grits.

Buffing plastics is usually done to restore the original lustre to the surface. A wide, flat cotton buffing wheel with Tripoli or other buffing agents applied to the surface, provides an effective polish. Care should be taken not to generate enough heat to soften the plastic. This is done by keeping the part on the buffing wheel moving at all times. A final buffing on a clean soft wheel with no compound will produce a good lustre.

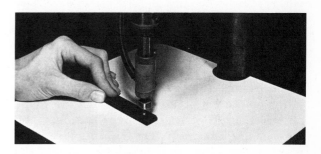

Fig. 17-49. Locating part under the hot staking head.

HOT STAKING PLASTICS

Hot staking is a mechanical fastening process used to fasten plastics to plastics and to other materials. It is similar to riveting metals except the head is shaped or spread by softening the plastic rather than by hammering. Most hot staking is done on molded products where the stakes are molded into the article. During assembly, the part with holes is placed over the part with stakes

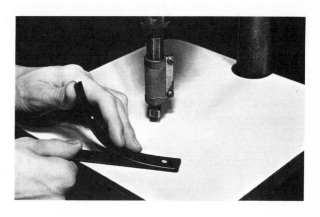

Fig. 17-50. Testing formed stake after cooling.

249

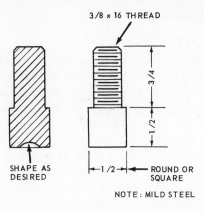

3/8 x 16 THREAD

3/4

1/2

SHAPE AS
DESIRED

1/2

ROUND OR
SQUARE

NOTE: MILD STEEL

Fig. 17-51. Working drawing of a hot stake attachment for a drill press.

and the halves are sealed as a unit. A hot tool melts the stakes and spreads them to form heads, Fig. 17-49. A rounded indentation on the hot staking tool forms the shape of the head. Hot staking provides fast and inexpensive assembly. The main disadvantage of using this fastening procedure is that the parts cannot be taken apart and reassembled.

Separate pieces of small diameter rod may be used as stakes and heads formed on each side of the parts. A hole is drilled through both pieces of plastic and the rod is inserted, extending about 1/8 in. from each side. Placed over a fixture with a hole in it of the correct depth, the part is hot staked on one side. The part is then turned over and lined up under the hot stake again. Light pressure is placed on the stake until it melts and spreads. After releasing the hot stake, the part is allowed to cool until rigid, Fig. 17-50.

Constructional details on a hot staking attachment for a drill press are shown in Fig. 17-51. The heating element and controls are the same as those illustrated in the section on hot stamping, page 255. Most of the thermoplastics can be hot staked, however, rigid materials like ABS and high impact polystyrene are most suitable.

HEAT SEALING PLASTICS

The basic theory of heat sealing is relatively simple. It involves the combining of two or more thicknesses of thermo-

plastic film or sheet by melting the layers together. However, the constantly changing techniques, equipment, and plastic films make it an area of concentrated study in the industry.

The major uses of heat sealing lie in the areas of packaging, Fig. 17-52, and the textile industry. A few examples of heat sealed products are inflatable toys, blister packages, sandwich wraps, automobile seats and convertible tops, raincoats, litter bags, billfolds and travel bags. The growth in demand for products such as these requires high-speed production rates with modern equipment and methods.

By definition, heat sealing is the process of welding specific areas of films together by using heat. It also applies to sealing films to other surfaces, such as cardboard, impregnated fabrics, laminated plastics, and foams. Most thermoplastic films can be heat sealed by one or more of the processes available. The major heat sealing processes are:
1. Thermal heat sealing.
2. Thermal impulse heat sealing.
3. Ultrasonic sealing.
4. Dielectric sealing.

THERMAL HEAT SEALING

Thermal sealing is the simplest and one of the most used heat sealing processes. It

Fig. 17-52. Heat sealed food package made from a combination polyethylene ionomer resin film.
(Du Pont Co.)

involves the use of heat, usually in the form of an electrically heated tool, bar, or die, to soften the layers of film to a molten state and fuse them together under pressure. The source of heat is continuous and cooling of the seal takes place after the heat and pressure are removed. A simple hand held heat sealer, Fig. 17-53, is similar to a thermostatically controlled soldering iron with a wide, flat base. This type of sealer is primarily used for packaging foods like cheeses and meats. The plastic film surrounding the food is lapped over and the flat face of the hot hand tool melts the layers of film together.

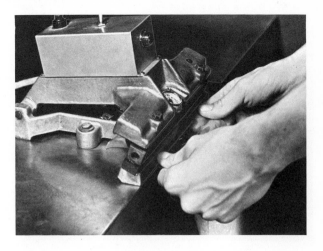

Fig. 17-55. Making the heat seal on a bar sealer.

Fig. 17-53. A hand operated thermal heat sealer for large surface sealing.

Bar type heat sealers, Fig. 17-54, are used extensively in hand packaging operations of foods and other products. For sealing, the films are placed between the hot

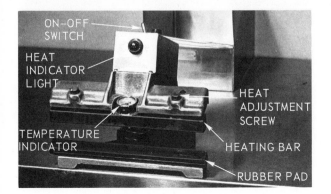

Fig. 17-54. A bar type heat sealer used in packaging. This sealer is foot pedal operated and is ideal for laboratory use.

upper bar and the lower pad, Fig. 17-55. Time and pressure for the seal are controlled by the operator using a foot pedal. Seals of this type must be watertight when packaging such foods as meats and pickles, Fig. 17-56. Automatic bar type sealers used

Fig. 17-56. Waterproof bags sealed on a thermal bar type sealer.

in industry for high-speed production operate on the same principle. Heated platen sealers are designed to seal various shapes, as shown in Fig. 17-57.

Straight line sealing of film for raincoats and bags may be done by hot rollers. The film is fed between the hot rolls, and a continuous seal is produced.

Fig. 17-57. A heated platen of circular steel discs seals the two layers of this air-trapped packaging film.

Fig. 17-59. Placing corrugated sheets with parts in place in a skin packaging machine.

Thermal heat sealing for skin and blister packaging is also in widespread use. In this process the film is softened by conduction or radiant heat and is sealed to a cardboard surface. Small particles of plastic resin are embedded in the surface fibers of corrugated cardboard, Fig. 17-58, to provide for the seal. Parts are located on the sheets and they are placed in the frame of the skin packaging machine, Fig. 17-59. The machine is actuated, turning on the heating element. As the plastic film is softened, the clamping frame lowers and drapes the film over the parts sheet. A vacuum is quickly drawn which pulls the soft sheet tightly around the parts and seals the film to the tray surface. As the film cools, the operator opens the film frame and pulls out the sealed tray which also pulls the film forward in the frame ready for the next cycle. The film is then slit to release the skin packaged sheet

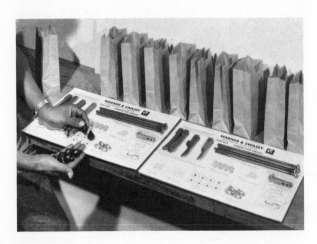

Fig. 17-58. Operator places parts on silk screened corrugated sheets which have small plastic particles embedded in the surface.   (Warner and Swasey Co.)

Fig. 17-60. Note how operator is carefully trimming finished sheet from large roll of polyethlyene film. Blister package sealing is done in a similar manner.

Fig. 17-61. A form-and-fill thermal heat sealing machine. Sealed packages are drawn off at the bottom. (Phillips Films Co.)

ribbon is embedded in a heat-resistant material to form the die. The bed which the die operates against is usually a rigid platen covered with a soft high-temperature material such as a silicone resin. The die is brought down, making contact with the films to be sealed against the bed. Upon contact, electric current heats the wire in the die to the fusion temperature of the plastic film. The seal is made and pressure held on the film as the die heating impulse is turned off automatically, Fig. 17-62. Cooling of the

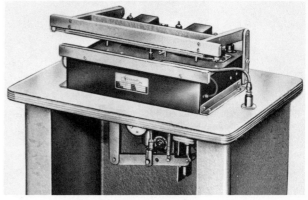

Fig. 17-62. Pneumatically powered high-speed thermal impulse heat sealer. (Vertrod Corp.)

from the roll of film, as shown in Fig. 17-60. Blister package sealing is done in a similar manner.

An automatic form-fill-seal machine for film packaging of products is shown in Fig. 17-61. This makes use of thermal sealing to form the bags. The continuous center-fold film is spread open, the product is inserted by the operator, and the film is sealed. The bag is also cut from the web of film, dropped to a product chute, and conveyed to the packing area. Thermal and impulse sealing are often used in combination in such packaging.

IMPULSE HEAT SEALING

Impulse sealing is similar to thermal bonding except the dies are hot only at the time of sealing. In this process, a wire or

seal is obtained under pressure as the die quickly cools below the fusion point of the plastic film. Impulse refers to the quick heating and cooling action of the die. The advantages of impulse sealing are the cooling of the seal under pressure, appearance, and lack of film distortion due to quick heating at the time of bonding. Impulse sealing finds a major use in film packaging, Fig. 17-62A. A portable impulse sealer suitable for laboratory work is shown in Fig. 17-63.

ULTRASONIC SEALING

The theory of ultrasonic sealing of film is the same as that of ultrasonic welding as discussed previously. The seal is made by mechanical vibrations of the sealing die against the plastic films. Heat is developed equal to the melting temperature of the film

Fig. 17-62A. The automated impulse heat sealer can bag at rates up to 45 packages per minute. (Automated Packaging Systems, Inc.)

by mechanical friction. As the seal is made, ultrasonic vibration is reduced and the bond is cooled under pressure of the die.

Continuous ultrasonic sealing is done by moving the film under a stationary tool. Ultrasonic sealing has the advantage of being able to seal some films which are not adaptable to other techniques. Ultrasonic equip-ment is considerably more expensive and operates at slower production speeds than most other forms of sealing.

DIELECTRIC SEALING

Dielectric or high frequency sealing involves the use of high frequency energy which is passed through the material to be sealed while it is held between a set of electrodes. As the plastic films resist the high frequency waves, their molecular structure is activated to the point that friction between molecules results in heat being generated. At the correct settings, enough heat is generated to melt the film and provide strong, uniform bonding. Fig. 17-64 shows a commercial dielectric heat sealing machine.

The equipment consists of a generator or frequency converter which converts low frequency power to high frequency electrical energy. This energy is supplied to the electrodes of a pneumatically operated press which raises and lowers the die to provide pressure for the seal. Finally, a die is required which will give the desired shape or outline of the seal.

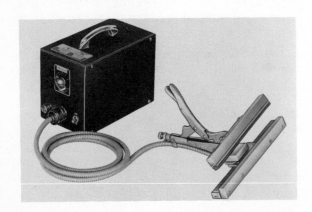

Fig. 17-63. A portable, hand type thermal impulse heat sealer.

Fig. 17-64. A commercial high frequency heat sealing machine with the frequency converter at the left. (J.A. Callanan Co.)

Fabricating Plastics Materials

The films are placed on the platen of the press and the die is lowered, exerting the correct pressure on the material. High frequency energy is then passed through the films for the required time cycle, the die is raised, and sealed parts are removed. The films most used in dielectric sealing, because of their favorable properties, are polyvinyl chloride, cellulosics, polyvinylidene chloride, and nylon.

MECHANICAL FASTENING

There are many fastening methods using plastics which provide good strength and high-speed production. These devices can be classified in two ways; integral or snap-fit fasteners, and external fasteners.

Snap-fit mechanical fastening relies on the flexible properties of plastics. Molded products are often designed in such a way that one part of an article will snap over and grip a part on another piece, Fig. 17-65. Such parts may have a loose fit so they may be taken apart or a tight fit for permanent assembly. A commercial assembly is shown in Fig. 17-66. Molded threads are also used for assembly of many products, Fig. 17-67. The threads eliminate the need for external fasteners and allow the part to be easily taken apart for repairs or replacement.

Fig. 17-66. Plastics are ideal for snap-fit container lids. Left. Lid is snapped shut. Right. Close-up of snap-fit detail. (Chemplex Co.)

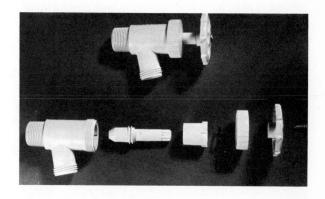

Fig. 17-67. Standard pipe threads are molded into the interchangeable parts of this acetal water valve for ease of assembly and repair. (Celanese Corp.)

Fig. 17-65. Snap-fit assembly of plastic parts which can be opened and closed many times. (Borg-Warner Chemicals)

External fasteners are used for the fabrication of many plastics products, either to fasten similar plastics or other materials. Standard bolts are used for the assembly of heavy sections as shown in Fig. 17-68. Self-tapping screws may be used in many plastics which are not too brittle to be cracked by the inserting pressure. Likewise, nails and staples are used in flexible materials for many construction purposes, Fig. 17-69. Metal and plastic spring clips which are forced over a stud also provide quick and inexpensive methods of fastening. Other de-

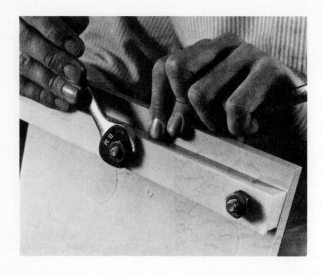

Fig. 17-68. Bolting heavy plastic sheet stock together.

Fig. 17-70. A rivet gun being used to insert rivets for sheet fabrication. (Borg-Warner Chemicals)

vices such as cotter pins, wood screws, and rivets, Fig. 17-70, find use in plastics fabrication.

Metal and plastic hinges are used on fabricated products. Plastic strip hinge material is available by the roll for many applications. This material, extruded from polyallomer or polypropylene, may be attached to most any material for hinge applications. In the laboratory, polyallomer

sheet made by the compression molding process may be cut into strips and strong hinges produced in the following manner:

1. Cut the polyallomer strip to the required dimensions.

2. Insert the wedge-shaped tool into the heating element on the drill press and set the temperature at 300 deg. F.

3. When the temperature has been reached, place the strip under the tool at the spot located for the hinge.

4. Bring the wedge-shaped tool down on the strip until it melts half way through the strip, Fig. 17-71.

Fig. 17-69. Stapling plastic sheet for construction panels. (Borg-Warner Chemicals)

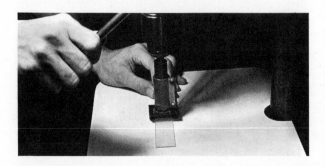

Fig. 17-71. Melting a V groove in a polyallomer strip.

Fig. 17-72. Flexing polyallomer strip while the V section is still hot.

and high density polyethylene will also serve for hinge purposes although they do not provide the flexing properties of polyallomer. Flexing the strip while the hinge section is still hot aligns the crystalline structure on an axis with the groove. Then the hinge may be flexed several thousand times without breaking.

5. Release the tool and quickly flex the strip two or three times while the hinge section is still hot, Fig. 17-72.

6. When the hinge is cool it is ready to be shaped for any application.

Construction details for a wedge-shaped tool are shown in Fig. 17-73. Polypropylene

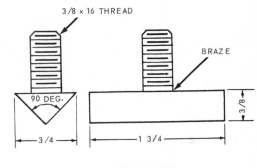

Fig. 17-73. Details of a wedge-shaped tool for making a plastic hinge.

TEST YOUR KNOWLEDGE - CHAPTER 17

1. The principle of bonding two plastic materials together may be classified as _____ _____ or _____.

2. Welding processes used with plastics are made up of four groups:
   a._____.
   b._____.
   c._____.
   d._____.

3. Hot gas welding makes use of heated _____ or _____ to soften the plastic to a molten state.

4. Plastics most often welded by the hot gas process are:
   a._____.
   b._____.
   c._____.

5. The name given to solvent cements which have a small amount of plastic dissolved in them are called _____ cements or _____ cements.

6. The most popular tips used in hot gas welding are the _____ tip, the _____ tip, and the _____ tip.

7. Four amorphous plastics easily bonded by solvent cements are:
   a._____.
   b._____.
   c._____.
   d._____.

8. Solvent cementing (does) or (does not) add a new material to the plastics being bonded.

9. Welding operations are used on:
   a. All plastic resins.
   b. Thermoplastic resins.
   c. Thermosetting resins.

10. Breathing the strong vapors of some solvent cements may cause _____ effects.

11. Hot plates used for fusion welding should be coated with _____ to keep the plastic from sticking to the surface.

12. Factors to be considered when selecting a welding process for a certain job are:
    a._____.
    b._____.
    c._____.
    d._____.

13. The most feasible friction welding technique is called _____.

14. How does speed welding differ from standard hot gas welding?

15. Ultrasonic welding produces about _____ cycles per second of mechanical vibration.

16. Why are soft plastics like polyethylene difficult to weld by the ultrasonic process?

17. The best hot gas welds require the correct:
    a. _____.
    b. _____.
    c. _____.
    d. _____.

18. Ultrasonic welding converts _____ to mechanical vibration.

19. Most of the machines and tools used for cutting and machining _____ and _____ are also used for plastics.

20. Plastic materials are usually cut by _____ or _____.

21. A _____ circular saw blade is usually the best for cutting plastic sheet.

22. Shearing is preferred over sawing plastic sheet because _____.

23. The major problem in machining plastics is _____ due to friction between the tool and the resin.

24. When tapping plastics, _____ taps allow more room for chip removal.

25. The rake angle on a lathe tool for turning plastics should be from _____ to _____.

26. Rigid plastic foam is usually cut by using a:
    a. _____.
    b. _____.
    c. _____.

27. Hot staking in plastics is similar to _____ of metals.

28. The main uses of heat sealing are in the areas of _____ and _____.

29. A main advantage of thermal impulse sealing is that _____.

30. Thermal sealing of plastic film to cardboard is accomplished by _____.

31. In thermal impulse sealing, the impulse refers to _____.

32. During continuous ultrasonic sealing operations:
    a. The tool moves.
    b. The film moves.
    c. Both film and tool move.
    d. None of the above.

33. Dielectric heat sealing is based on the principle:
    a. Of inducing external heat to make the seal.
    b. Of mechanical vibration to melt the film.
    c. Of intermolecular friction to soften the film.
    d. All of the above.
    e. None of the above.

34. The properties of the following films make them well suited to dielectric heat sealing.
    a. _____.
    b. _____.
    c. _____.
    d. _____.

35. Mechanical fasteners may be classified in two ways, as _____ fasteners and as _____ fasteners.

36. _____ possesses the best properties for making plastic hinges.

37. Plastic films are often heat sealed to other materials such as:
    a. _____.
    b. _____.
    c. _____.
    d. _____.

38. During hot staking, a hot tool _____ the stake and _____ it to form the head.

39. A _____ file is used to remove deep tool marks from plastics while a _____ file is used for smoother finishing.

40. When turning plastics, a _____ and a _____ will keep from softening the plastic by frictional heat.

SUGGESTED ACTIVITIES

1. Using a heat sealer, make a watertight packaging bag from plastic film.

2. Develop a research project to check the effects of solvent cements on a number

of different plastic resins. Try a drop of each solvent on the surface of each resin. Record the results. Try to determine why the structure of each polymer may cause a different reaction to the solvents.

3. Visit a plastics plant that does plastic welding. Ask what materials and techniques they use. Describe the processes to your class.

4. Develop a series of posters on the topic of safety tips when welding plastics.

5. Make a collection of mechanical fasteners used for plastics fabrication and develop a bulletin board display around them.

6. Develop a paper, with illustrations if possible, on the early attempts to heat seal plastic film.

7. Design and construct a fixture to hold the end of a plastic rod while doing a hot staking operation.

8. Make a collection of plastic films that have been heat sealed. Try to determine what method of heat sealing was used for each article.

9. Demonstrate to your class the proper way to turn plastics in a lathe. Grind a tool bit and explain why the tool must not have a positive rake angle.

10. Set up an experiment to impregnate the surface of a piece of cardboard with powdered polyethylene. Try heat sealing a polyethylene skin package to the surface.

11. Prepare a display that shows the different uses of plastic welding on product fabrication.

12. Write to manufacturers and suppliers for literature on solvent cements. From the material you receive, make a chart listing thermoplastic resins and a recommended solvent cement for each.

13. Make a diagram showing the principle of ultrasonic welding. Ask your school's electronics instructor for information concerning symbols and the layout to be used. Write to manufacturers for specifications on their equipment.

This is a production four color rotogravure printing press for polyvinyl chloride film. (Inta-Roto, Inc.)

Note the many plastics products which require decorating and finishing.

Chapter 18
DECORATING AND
FINISHING PLASTICS

Included in decorating and finishing are a great variety of processes by which plastics articles are painted, coated, colored, textured, and enhanced with designs and messages. See Fig. 18-1. Plastic products carry many types of communication, from soup packages, Fig. 18-2, to a decorative wrapping, Fig. 18-3. In the plastics industry most of the decorating or finishing is done during molding, directly afterward, or prior to assembly.

VACUUM METALLIZING PLASTICS

Appealing finishes for many plastics products may be obtained by using a process known as vacuum metallizing. This consists of coating plastics with a thin layer of metal

Fig. 18-2. Plastic bags of ionomer film require special decorating techniques.
(Du Pont Co.)

which provides a bright, attractive, metallic appearance, Fig. 18-4. Metallizing provides a coating that will withstand moderate wear.

Most plastics can be satisfactorily vacuum coated, although polystyrene, polyester, polycarbonate, phenolic, and ABS are most frequently used. Some polymers require special surface treatment prior to coating. The selection depends primarily on product use and cost. Fig. 18-5 illustrates metallizing of inexpensive products made from polystyrene.

The vacuum metallizing process requires a series of relatively simple steps. The parts are first given a base coat of lacquer by spraying, Fig. 18-6, or dipping to even

Fig. 18-1. Most appliances like this microwave oven require extensive decorating, coloring and texturing.
(General Electric Co.)

Fig. 18-3. Colorful decorating of plastic film used for bread wrappers. (Dow Chemical Co.)

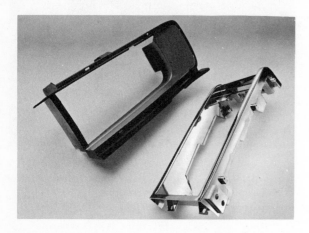

Fig. 18-4. Vacuum metallizing is widely used to provide metallic surfaces on automotive dash components. (Dow Chemical Co.)

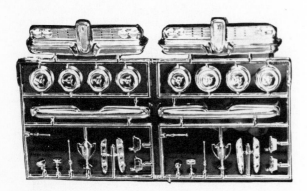

Fig. 18-5. Brilliant metallic looking parts are vacuum metallized on injection molded polystyrene toy kits.

Fig. 18-6. Vacuum metallized parts require two coats of lacquer, applied and baked before and after metallizing. Here a conveyor is loaded ready to carry the parts through the spraying and baking areas before the metallizing operation. (Michigan Oven Co.)

out the surface defects which would be emphasized by the brilliant metallic finish. After oven drying, the parts are placed on the rotating racks of a portable carriage. The carriage contains many tungsten wire filaments, heated electrically, to which small pieces of aluminum wire are attached. Next, the carriage is placed in a vacuum chamber,

Fig. 18-7. Vacuum metallized parts on the carriage rack inside vacuum chamber.

Decorating and Finishing Plastics

Fig. 18-7, and the door is tightly sealed. A vacuum pump is used to draw out as much air as possible. A transformer is then turned on to heat the tungsten filaments which causes the aluminum strips to vaporize. As the aluminum vaporizes, the molecules (aluminum vapor) drift directly to the cool plastic parts and condense on the surface. The plastic parts are rotated so all surfaces are evenly coated.

After the coated parts are removed from the chamber, they are sprayed with a protective coating of lacquer. The aluminum surface is easily scratched and the tough lacquer coating provides the needed protection.

The aluminum coating deposited on the plastic surface is extremely thin, about 5 millionths of an inch in thickness, and is used mostly for decorative purposes rather than for protection or abrasion resistance. Plastic film for gift wrapping and packaging is metallized in a similar manner by using a continuous roll process.

ELECTROPLATING PLASTICS

Plastics are nonconductors of electricity but they may be electroplated if the surface is treated to receive a metallic deposit. The equipment used is essentially the same as that used in any standard electroplating system. The advantages of plating plastics with metal are many. Parts may have the advantage of plastics properties and yet retain their metallic applications. This is true in the plumbing industry, Fig. 18-8, and in small appliances. Many metal parts may be replaced by lightweight plastics and still serve the desired purpose. Typical applications include housewares; automotive parts such as tail lamp housings and grills, door handles, and nameplates. See Fig. 18-9. Each combines the light weight and corrosion resistance of plastics with the wear resistance and sparkle of a metal surface.

Fig. 18-9. Chrome plated auto trunk assist handle, exterior mirror and window crank combine beauty with light weight and durability. (Allied Chemical Co.)

Fig. 18-8. Chrome plated plastic faucets and plumbing parts are in increasing demand.

Commercially, two plating processes are currently being used. The first involves the application of a conductive film, usually silver, in a silver nitrate solution. The part is then copper plated with a thin coating in a bath containing copper anodes. Finally, nickel or chrome plating finishes are applied, Fig. 18-9A.

A process called electroless plating, produces a copper coating on the plastic surface quicker and with less expense than the system described in the preceding paragraph. This consists of immersing the plastic part in a bath of a metal salt, then a solution of a copper or nickel salt, and normal electroplating with chrome or nickel. Plastics used in plating systems include ABS, polycarbonate, acetal, phenolic, and urea.

Fig. 18-10. Note hot stamped marking on acrylic decorative plates. (Gladen Div., Hayes-Albion Corp.)

Fig. 18-9A. Chrome plated plastics products are being removed from electroplating bath. (Allied Chemical Co.)

HOT STAMPING

Marking and decorating plastics by the hot stamping process has been done for many years. Recent developments in equipment and hot stamping foils, however, have made it one of the more popular decorating techniques in the industry. See Fig. 18-10.

The basic hot stamping process consists of transferring a design or pattern from a thin polyester or cellophane film backing (often called a foil) onto the plastic product. Heat and pressure against the back of the foil make the design adhere to the plastic product.

Hot stamping equipment ranges from high-speed production units, Fig. 18-11, to smaller laboratory models as shown in Fig. 18-12. Fixtures to support the product and appropriate dies make it possible to stamp on almost any surface.

Foils utilized for the stamping process usually consist of a special pigmented or metallized coating on a carrier film. The pigment or metallized coating should be compatible with the plastic to be stamped, because it must fuse and adhere to the plastic surface. Foil manufacturers product many different formulations for specific resins. Foils are available in many degrees of abrasion resistance, chemical resistance, as well as in a wide range of colors. Specialty foils are also produced for specific applications. Look at Fig. 18-13. This variation in the process is commonly known as heat transfer decoration.

Fig. 18-11. Automatic hot stamping design on plastic tumbler with
production unit.   (Acromark Co.)

HOT STAMPING DIES

Metal dies for hot stamping are usually made of steel, magnesium, brass, or zinc. Magnesium and zinc are chemically etched to produce the die design and are normally

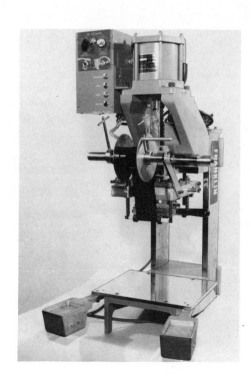

Fig. 18-12. A versatile laboratory size hot stamping machine.
(Franklin Imprinting Machines)

Fig. 18-13. A multicolored heat transfer foil on a paper backing. The part is hot stamped in four colors in one operation.
(Gladen Div., Hayes-Albion Corp.)

used for short production runs. Brass and steel dies are usually machined or engraved. Steel dies are used for extremely long production runs.

Silicone resin dies are especially popular because of their flexibility, low cost and adaptability to irregular surfaces. As a flat pad, they are used to hot stamp raised letters, designs, and edges. Molded silicone dies may be used similar to metal dies, Fig. 18-14. Textured flat or roller dies, Fig. 18-15, can be used for continuous design applications. Large surface areas to be hot stamped are making use of heated silicone

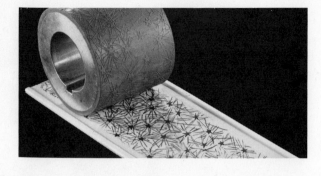

Fig. 18-15. Textured silicone hot stamping rolls produce continuous designs.

Fig. 18-14. Molded silicone hot stamping dies fused to an aluminum backing. (Gladen Div., Hayes-Albion Corp.)

Fig. 18-16. Hot, four-roll silicone dies transfer wood-grained decoration to the flat sides of this television cabinet. The automatic hot stamping machine raises the cabinet into position for the rollers to move across on all sides. (Service Tectonics, Inc.)

rollers to transfer the design, Fig. 18-16. A newer technique in die production combines the hardness and wear of steel with the flexibility of a silicone resin. These dies are known as sandwich dies, Fig. 18-17.

HOT STAMPING PROCESS

The hot stamping operation involves placing the piece to be decorated in a stamping machine using a fixture to hold it in place. The desired color of pigmented foil is placed between the hot die and the plastic piece. The foil which is in roll form advances

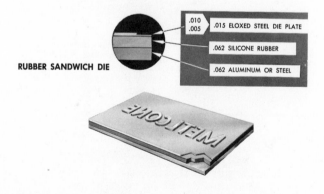

Fig. 18-17. A sandwich die with a thin steel stamping face and a flexible silicone backing makes good transfer on non-uniform molded surfaces. (Gladen Div., Hayes-Albion Corp.)

during each cycle to place an unused portion under the die. As the die lowers under pressure, it is held against the product a predetermined time. During the pressure period, the pigment from the foil transfers to the plastic surface and fuses in place. The die returns to its open position and the foil is pulled away from the part. This is the process used for hot stamping most plastics products. See Fig. 18-17A.

Laboratory hot stamping can be done on a small commercial machine or by adapting a drill press to the job by fitting it with a

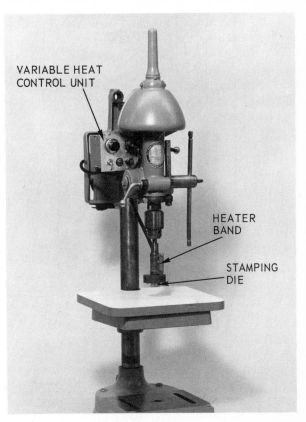

Fig. 18-18. Heating unit for hot stamping is held in the drill press chuck. A temperature control device is mounted at the side of the press.

Fig. 18-17A. The design for these containers was printed in reverse on clear film and then hot stamped on the part. (Dennison Manufacturing Co.)

heating element and controls as shown in Fig. 18-18. The same unit was used for hot staking and making the groove for plastic hinges. Details of a heating element for interchangeable type letters is shown in Fig. 18-19. The procedure for hot stamping a design on a typical project, such as injection molded, low density polyethylene bottle caps is as follows:

1. The die is made up using metal type of suitable size. The die unit is attached to the drill press chuck, and the proper foil is selected for the operation. See Fig. 18-20.

2. Heat is turned on and the die is allowed to warm up. The exact temperature may be determined with a portable pyrometer, Fig. 18-21. A temperature of about 200 deg. F. is required. If a pyrometer is not readily available the proper temperature can be determined by making tests.

3. The bottle cap is located on a fixture to hold it in stamping position, Fig. 18-22, and a piece of foil placed over the part with the pigment side down.

4. The die is brought down to contact the part with light pressure, held momentarily, and released, Fig. 18-23.

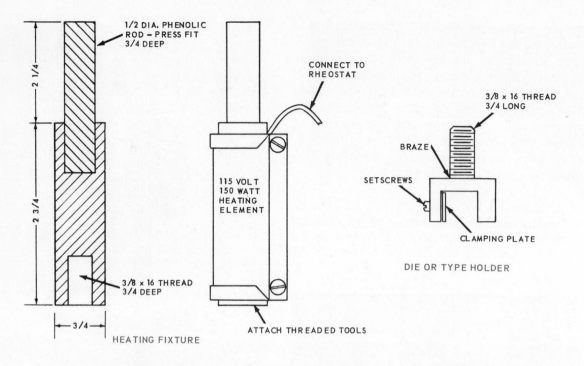

Fig. 18-19. Left. Heating fixture for threaded tools. Right. Die holder for hot stamping. Make dimensions appropriate for die or type being used.

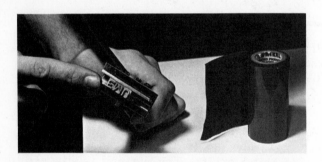

Fig. 18-20. Printing type set in the die holder and a roll of foil to be used in hot stamping is selected.

Fig. 18-21. Checking die temperature with a portable pyrometer. Die temperatures are determined from foil manufacturer's specifications.

5. The foil usually sticks to the part and must be carefully pulled away, Fig. 18-24.

6. The cap is removed from the fixture and another cap inserted while the foil is moved to an unused portion.

Small parts hot stamped on a high-speed

Fig. 18-22. Locating the bottle cap on the hot stamping fixture. A center plug under the cap supports the surface to be stamped.

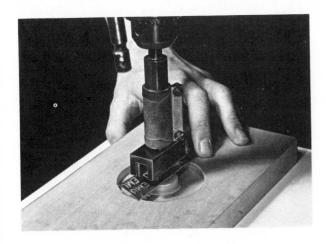

Fig. 18-23. Applying pressure to the foil covered cap.

production basis are shown in Fig. 18-25.

Fig. 18-25. Special tooling is mounted on a turntable for hot stamping three sides of a small game piece. (Kensol-Olsenmark, Inc.)

A useful tool for laboratory production and repair work is a hand hot stamping unit, Fig. 18-26. It is used for decorating raised surfaces molded into a product, edges of parts, and small flat surfaces. The tool is set to the desired temperature and is moved slowly across the foil which has been placed over the part.

SILK SCREEN PRINTING

The silk screen process for labeling and decorating plastics products has been used successfully in the industry for many years. This provides very good wear resistance as well as an inexpensive application. Silk screening is generally used for precision printing and intricate line work rather than for covering large surfaces.

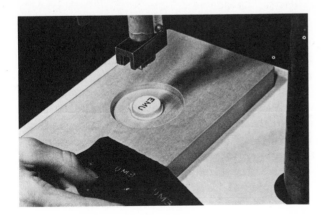

Fig. 18-24. The foil is stripped from the cap leaving the transferred pigment.

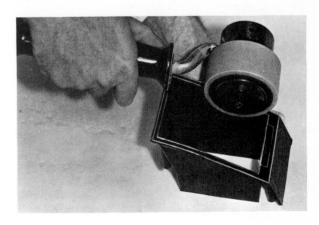

Fig. 18-26. A hand hot stamping tool with a rheostat controlled, electrically heated silicone roller. (Service Tectonics, Inc.)

The silk screen process consists of transferring an ink or paint through the openings of a fine screen to the product surface. A hand or mechanical squeegee is used to force the ink through the design in the screen, Fig. 18-27. Although the term "silk screen" is used, metal and plastic screens (especially nylon) are generally used with plastics. The blank screen has a water or chemical solvent coating, and a stencil, usually photographic, is used to produce the desired design. Intricate line work is especially adaptable to photographic stencils.

Fig. 18-28. This single station screen printer, by making use of an automatic part-flipper, decorates all four sides of a beverage case. (American Screen Process Equipment Co.)

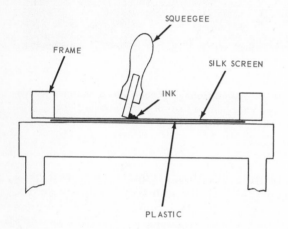

Fig. 18-27. Drawing which illustrates principle of silk screen process. The printing medium (ink, paint) is forced by a squeegee through the fine mesh of a screen stretched over a printing frame to form design on plastic surface.

In both hand and mechanical printing, Fig. 18-28, the plastic part is placed on a fixture about 1/8 in. below the screen. As the squeegee passes across the screen, the printing ink is transferred to the part through the openings. The screen is lifted and the printed plastic part removed.

Numerous inks and paints are available for printing on practically all plastic surfaces. The ink adheres to the plastic surface by solvent action and is usually dried very quickly by passing the products through an oven. Inks that cure through polymerization usually provide a tougher, more scratch resistant surface.

Production silk screen printing is usually done on automatic equipment as shown in Fig. 18-29. Large production units are normally designed to handle almost all product shapes including bottles, plates, and film or sheet.

IN-MOLD DECORATING

Many plastics products are decorated by the in-mold process. This consists of placing a printed plastic overlay film in the mold before final molding of the product. The decorated film fuses into, and becomes an integral part of the molded piece. Decorations of this type are durable and permanent, Fig. 18-30.

In-mold decoration is generally used in compression molding of thermosetting resins, and in injection molding of thermoplastic materials. Decorating thermosetting materials consists of loading and closing the mold in the conventional manner. When resin has partially cured, the mold is opened and the overlay is positioned in the mold. The mold is then closed and molding proceeds as in general practice to final cure. The overlay is made from a printed cellulose sheet covered with a partially cured melamine film. After molding, the melamine film makes an integral bond with the product.

Fig. 18-29. A fully automatic screen printing line for plastic containers. It can print up to 3,600 items per hour. (American Screen Process Equipment Co.)

A similar process is used in injection molding except it is a one-step operation. The overlay is placed in the mold prior to injection of the resin. In this case, the overlay is made from a printed film of the same polymer as the product so a good bond results. To keep the overlay from shifting when the injection shot is made, the overlay is usually charged with static electricity.

Fig. 18-30. This melamine dinnerware for children is decorated by the in-mold decorating process. (Lexington United Corp.)

PAINTING PLASTICS

Although plastics products are easily molded in a wide range of colors, painting is often necessary to achieve the desired decorative effect. This is particularly true when two or more colors are required on the same article. All plastics can be painted, but some resins do not provide a good surface adhesion as required for lasting results. Major considerations when painting plastics are the compatibility of solvents in the paint with the plastic surface, adhesion of the paint film, coverability, and resistance to wear and chemicals. Such products as automobile dash panels, television and radio cabinets, toys, and home appliances, make use of a number of painting operations.

Fill-in painting is a technique used on many products which have a design molded in the surface. Paint, of proper consistency and containing an appropriate solvent, is applied by spray or by hand, Fig. 18-31. This method is particularly effective on dials and knobs where surface wear does not remove the design.

Fig. 18-31. Fill-in painting used on a spinning reel molded from acetal resin. (Celanese Plastics Co.)

Spray painting has become a valuable process in the decoration of plastics. Some molded products are completely finished by hand spraying. Automatic spray painting is extensively used for high-speed production work. Most systems use some type of a masked spray painting unit in which a mask covers the parts of the product which are not to be painted as it passes by the spray. The mask is usually made of metal and conforms to the contour of the surfaces to be protected from the spray. Parts to be painted are conveyed from the molding area, to automatic spray painting machines, Fig. 18-32. Here they are placed on fixtures

which locate parts under the mask, Fig. 18-33. The parts are raised to seal against the mask and fed under spraying unit. After painting, they are air or oven dried.

Another painting method is shown in Fig. 18-34. In this system, large rollers made of soft rubber continuously receive a stream of paint. By rotating together the paint is evenly spread over the roller surfaces. The molded product to be painted is

Fig. 18-33. The operator loads dash panel parts on the spraying fixture which will rotate to the rear of the machine. The spray head has just moved across a set of panels previously loaded.

Fig. 18-32. A conveyor line carrying molded automotive dash panels to spray painting machines.

Fig. 18-34. Roll painting the flat face and edges of an automobile dash panel.

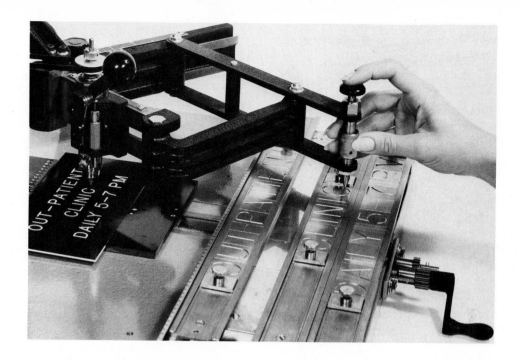

Fig. 18-35. Using engraving machine to cut letters in laminated plastic sheet.
(New Hermes Corp.)

placed in a fixture and is passed along the roller surface. Only exposed flat and irregular surfaces can be roller painted.

ENGRAVING PLASTICS

Mechanical engraving of laminated plastic sheet provides a permanent method of marking and decorating. It is used extensively for the production of nameplates, dials, tags, and signs. Laminated sheets may be purchased in various colors and thicknesses, or they can be made in the laboratory. The laminate consists of two or more plastic sheets of different colors sealed together. See page 191. An engraving machine traces over metal patterns, Fig. 18-35. The engraving tool cuts through the top plastic surface. The second layer of colored plastic is then exposed to show the design or letters. Designs, using an engraving machine, may also be cut into solid plastic or molded parts.

TEST YOUR KNOWLEDGE - CHAPTER 18

1. Vacuum metallizing plastic parts makes use of _____ for the coating material.
2. The vacuum metallized coating over a plastics part has a thickness of about:
   a. 10 thousandths of an inch.
   b. 1 millionth of an inch.
   c. 5 millionths of an inch.
   d. None of the above.
3. List three advantages of plating plastics with a metal coating.
   a. _____.
   b. _____.
   c. _____.
4. Since plastics are nonconductors of electricity, it is necessary to coat the surface with a _____ before plating.
5. Metallizing is one of the most _____

plastics finishing techniques.

6. The most commonly used plastics for vacuum metallizing are:
   a._____.
   b._____.
   c._____.
   d._____.
   e._____.

7. The wearability of vacuum metallized surfaces is:
   a. Poor.
   b. Moderate.
   c. Excellent.
   d. None of the above.

8. The parts in the vacuum metallizing chamber must be continually rotated because _____.

9. Abrasion resistance is provided for the surface of metallized parts by_____.

10. Foils used for hot stamping consist of a _____ or_____ coating on a_____.

11. Metal dies for the hot stamping process are usually made of:
    a._____ .
    b._____ .
    c._____ .
    d._____ .

12. _____ is used for hot stamping dies for long production runs.

13. Heated _____ _____ are used to transfer designs from the foil to large surface areas.

14. Silicone hot stamping dies have the advantages of being:
    a._____ .

b._____ .
c._____ .

15. Pigment transferred to a plastic part is fairly permanent because it _____ to the surface.

16. The advantages of silk screen printing on plastics are _____ and _____.

17. Explain why in-mold decoration of plastic products is so permanent.

18. To keep the overlay from shifting during in-mold decorating it is _____.

19. The main advantage of fill-in painting is_____.

20. Four considerations to be taken into account when painting plastics are:
    a._____.
    b._____.
    c._____.
    d._____.

21. A _____is used to keep sprayed paint off of areas of the plastic part not to be coated.

22. Mechanical engraving is used extensively for making:
    a._____ .
    b._____ .
    c._____ .
    d._____ .

23. The screen used in silk screen printing of plastics is usually made of_____ or_____.

24. A hand hot stamping unit is used to decorate:
    a._____.
    b._____.
    c._____.

SUGGESTED ACTIVITIES

1. Make a drawing which will illustrate the vacuum metalizing process. Write to manufacturers for literature and request pieces of the metal used to coat plastic products. Attach these to your drawing.

2. Design and construct a fixture to hold a plastic product for hot stamping on your laboratory press.

3. Construct a small silk screen unit to print a selected design on a production plastic part. A vacuum formed product is often suitable for this decorating activity.

4. Make a display board that features products that have been hot stamped for decorative purposes. Write to manufacturers of hot stamping foils and request samples or collect samples from your laboratory. Use these to illustrate the transfer process on your display board.

5. Using your laboratory engraving machine, make a nameplate for a machine

or cabinet in your laboratory.

6. Prepare a written report on the most recent developments in plating plastics. Use your technical library and ask your instructor for resource material.

7. Secure samples of plastics that have been electroplated to show to your class.

8. Prepare a demonstration of the hot stamping process.

9. Make a drawing of a silicone die used for hot stamping. Show the die in cross section and label the parts.

10. Prepare a paper that will explain the different types of masks used in automatic spray painting of plastic parts. Include the method used in the system to clean the masks.

11. Secure literature on various hot stamping machines presently used in industry.

12. Write to manufacturers requesting samples of overlay films used for in-mold decorating. Make a display of these overlays which shows their application.

13. Construct a storage container for rolls of hot stamping foil.

14. Make a silicone hot stamping die using carving tools and blank silicone die stock. Test the die on your press.

Production decorating of glass bottles with a protective plastic coating reduces problems from broken glass.
(E. I. du Pont de Nemours and Co., Inc.)

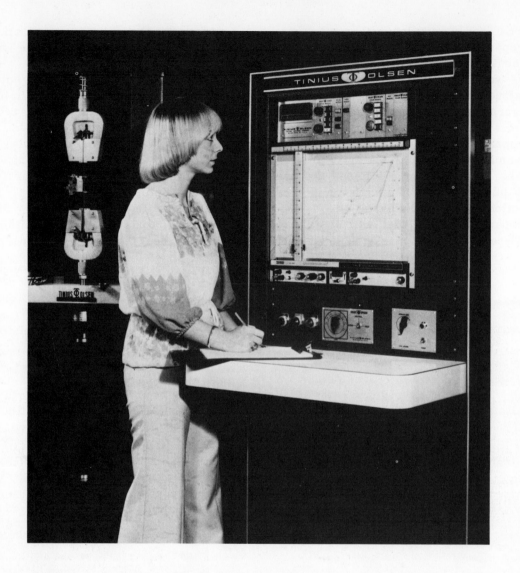

This plastics laboratory testing equipment operator would be classified as a technician.
(Tinius Olsen Testing Machine Co.)

Chapter 19
OPPORTUNITIES IN PLASTICS CAREERS

The plastics industry is expanding at such a rapid pace that opportunities for careers in this field appear to be almost unlimited. New and varied developments in plastics processing continue to add to the long list of jobs requiring the services of people interested, educated, and dedicated to the field. In the future, many new jobs will be created, greater services will be required, and employees educated in many phases of plastics will be a necessity.

A person interested in employment in any area of the plastics industry should have a fundamental knowledge of plastics materials and processes. Since the industry is so varied, from the securing of raw materials to the final product, an understanding of broad job classifications is desirable.

Job classifications in the plastics industry fall under these main categories:
1. Semiskilled.
2. Skilled.
3. Technical.
4. Supervisory.
5. Professional.

SEMISKILLED WORKERS

Workers whose jobs do not require a long period of training or skill are classified as semiskilled. Only a few hours of instruction may prepare the semiskilled worker in the plastics industry for the jobs to be performed. Typical assignments of the semiskilled worker may be generally classified as follows:

1. Persons who operate semiautomatic or automatic molding machines.
2. Persons who assemble plastics components into the final product.
3. Persons who transport molding materials, products, and packaged shipments.
4. Persons who handle various plastics materials - - bagging, mixing, blending, etc.
5. Those who assist skilled workers with inspection, machine maintenance, and repair work.

More persons are employed in the plastics field as semiskilled workers than any other group. This is primarily due to the vast production practices and assembly line operations within the industry. Resourceful persons may well advance to other positions within the industry by furthering their education in specific technical training programs.

SKILLED WORKERS

The classification of skilled workers in the plastics industry is a smaller group than in many other fields of manufacturing. This is due to the limited amount of hand work done in plastics processing. Skilled workers include mold builders, Fig. 19-1, machine service workers, machine setup persons and general machinists. Most skilled workers in the plastics industry are machinists who have obtained their skills through an apprenticeship or trade training program. Further education dealing with plastics materials and processes is a necessity. This may be obtained through on-the-job training programs and in technical schools.

Fig. 19-1. Plastics mold development requires the skills of a tooling specialist and mold builder. (The Budd Co.)

The machine setup person is one of the most important in the area of skilled workers. It is the setup persons' responsibility to mount the molds, adjust dies, and make trial runs before the machine goes into production. They are usually skilled machinists with a good background in plastics materials and molding science. They should have a knowledge of print reading and mathematics. The setup person often trains the machine operator in the correct use of the machine and how to check the quality of the product.

TECHNICIANS

The technicians play a broad role in the plastics industry. They are the connecting link between the skilled worker and the engineer. In some cases, their responsibilities are far reaching, but most technicians are specialists in some phase of the industry. The technician must have a knowledge of the total industry which usually requires at least two years of college with emphasis on math, science, manufacturing processes, and plastics technology. A person interested in becoming a plastics technician should inquire into the available programs on plastics technology at community colleges, vocational schools, and universities.

Technicians often work in product design and development along with the engineering staff. In some companies, they are responsible for making technical reports, checking production machine operation, making repairs on precision equipment, and a variety of testing and quality control work. See Fig. 19-2.

SUPERVISORS

Supervisory positions in the plastics industry are those that deal primarily with production problems and personnel. Supervisors are often plant managers or produc-

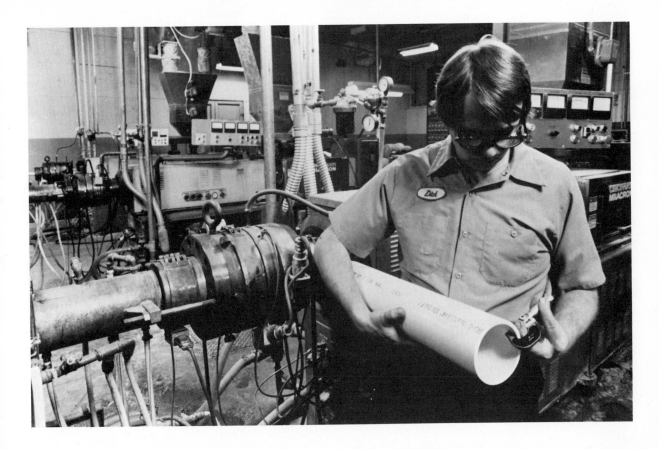

Fig. 19-2. The plastics technician in this extrusion facility is responsible for testing and quality control of extruded products.   (Cincinnati Milacron Co.)

tion managers. They are responsible to the plant management for meeting production schedules, maintaining molding material inventories, checking and supervising machine operators, and meeting shipping schedules, Fig. 19-2A. Plant supervisors check not only the parts but are also responsible for monitoring press performance.

The supervisor is required to have a general knowledge of all parts of the production system. An in-depth knowledge of all molding processes as well as experience in industrial production work is desirable. Supervisory positions are usually obtained by persons with a number of years of production experience and a broad technical background. At least two years of college in manufacturing processes with emphasis on plastics processing, provides good preparation for this type of position.

Fig. 19-2A. Plant supervisor is checking part and press performance.

PROFESSIONAL PERSONNEL

Professional people are in great demand in the plastics industry and the opportunities are rewarding. Professions in the field of plastics may be generally classified as those requiring at least a four year college education. These include the chemical and mechanical engineer, the research chemist, sales personnel, and the educator.

Chemical engineers usually assume responsibilities dealing with the physical properties of the plastics compounds used by their company. They work with the research chemist and engineering staff on problems concerning material compounding, mixing, and blending for particular products. They are concerned with machinery and mold design as related to the flow patterns and quality of the molded product. In many cases, the chemical engineer experiments with new plastics materials by adding fillers and plasticizers. Much of their work is of a laboratory nature.

A person working as a mechanical engineer in the plastics industry is usually responsible for machinery and tool design, development of new molds and dies and planning of production facilities, Fig. 19-3. They often supervise the installation of new equipment and plan production schedules.

The research chemist must, of course, have a good knowledge of organic chemistry. The large chemical companies and producers of plastics resins need qualified chemists, usually requiring a B.S., M.S., or Ph.D. degree. Many manufacturing companies who produce their own resins also employ research chemists. Their responsibilities include research and investigation into the development of new polymers, analysis of polymer structure and testing of plastics properties for the continual demands of modern industrial applications.

Plastics sales personnel are divided into two groups: those whole sell for the resin manufacturer and those who sell the services of custom molders. In each case, they must

have a broad knowledge of plastics materials and the processing industry. They should preferably have a college degree in sales engineering or business and should receive in-service training on the particular products they are expected to sell. They should possess a pleasing personality, enjoy selling, and be willing to travel as their job demands.

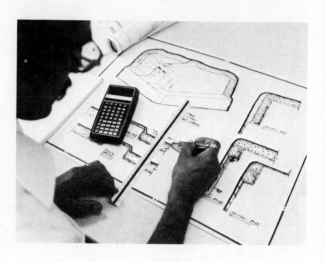

Fig. 19-3. A mechanical engineer may specialize in plastics product design which requires a great deal of science and math.   (Owens-Corning Fiberglas Corp.)

There is a great need for plastics education at all levels: for the public, for those preparing to enter the industry, and for many already engaged in the industry. Teaching is a rewarding and satisfying profession that many young people should consider as they anticipate their future careers. Teachers with a strong background in plastics are needed in the high schools, technical schools, and universities. Four years of college education are necessary and occupational experience is quite helpful to prepare for a teaching career.

Opportunities in plastics careers are as varied as in any industry. Teams of employees, Fig. 19-4, at every level of the plastics industry are constantly engaged in some of the most challenging and exciting activities in the world of work. Since the

Fig. 19-4. The mechanical engineer, machine designers and plant manager (employee team) are discussing aspects of a new molding design system.   (Cincinnati Milacron Co.)

plastics industry is expanding at such a rapid pace, opportunities for careers in this field appear to be almost unlimited.

OCCUPATIONAL INFORMATION ON PLASTICS

Many sources of information are available on careers in the plastics industry. Your school guidance department and industrial education teachers are readily available, as is material from the school library.

Industrial organizations such as the Society of the Plastics Industry and The Society of Plastics Engineers provide literature and information concerning job opportunities, educational requirements, and educational programs.

Information on plastics courses and educational programs may be secured from most community or junior colleges and universities. Many of these institutions offer technical programs in plastics.

TEST YOUR KNOWLEDGE - CHAPTER 19

1. More people are employed in the plastics industry as _____ workers than any other group.
2. The two groups into which sales personnel in plastics are usually divided are:
   a. _____.
   b. _____.
3. What five general categories are listed

for plastics occupations?

a._____.

b._____.

c._____.

d._____.

e._____.

4. The _____ is one of the most important of the group of skilled workers.

5. Three responsibilities of the research chemist are:

a._____.

b._____.

c._____.

6. List three sources of information on careers in the plastics industry.

a._____.

b._____.

c._____.

7. The mechanical engineer in the plastics industry is often responsible for:

a. Planning production facilities.

b. Development of new molds.

c. Tool and machinery design.

d. All of the above.

e. None of the above.

8. Education for professions in the plastics industry usually requires: _____.

9. A person who is often the link between the skilled worker and the engineer is the _____.

10. The_____often works on problems of mixing, compounding, and blending plastics materials for specified products.

SUGGESTED ACTIVITIES

1. Contact the Society of the Plastics Industry for literature and information on educational programs to prepare people for jobs in the plastics industry.

2. Make a list of job classifications and titles from materials in your school library. Try to find out what educational requirements are necessary for each job classification in the plastics industry.

3. Visit a local plastics company and discuss job opportunities with their personnel director. Make a report to the class on the job classifications of the people they employ, the necessary educational requirements, and the tasks they perform.

4. Make a bulletin board display around the topic, "Careers in Plastics." Use pictures from magazines and industrial literature to depict the variety of job opportunities.

5. Prepare a report on the job responsibilities of people employed in the resin manufacturing industries. Write to chemical companies for literature and use your technical library for resource material.

6. Select films dealing with plastics processing from your school film catalog. Make arrangements to barrow films in order to see what is required of persons engaged in processing.

7. Look through the employment columns of a number of newspapers and prepare a listing of the opportunities available in plastics or related jobs. Place each job opportunity under the general job classifications. Note which job classification appears to be in need of the most employees.

REFERENCE SECTION

SHEET MATERIALS FOR THERMOFORMING

RESIN SHEET	FORMING TEMPERATURE	FORMABILITY	COLOR AND CLARITY	SPECIAL CONSIDERATIONS
Acrylic	260 to 360°F.	Good	Brilliant clarity, all colors from opaque to transparent.	Heat should be applied from both sides. Even stretching, good for free blowing.
Cellulose Acetate Butyrate	265 to 320°F.	Excellent	Good clarity, good color range.	Deep draws well, heating should be fast.
Acrylonitrile Butadiene Styrene	300 to 350°F.	Good	Good range of opaque colors.	Excellent impact strength, average forming characteristics.
Polyethylene	325 to 425°F.	Fair	Naturally milky, many colors available.	Good toughness, sags easily when forming, best for shallow draws.
Cellulose Acetate	270 to 325°F.	Excellent	Good clarity and range of colors.	Needs fast heating, blisters easily, shallow drawing.
Polyvinyl Chloride	255 to 355°F.	Excellent	Transparent to opaque in many colors.	Good deep drawing, high impact strength.
High Impact Polystyrene	365 to 385°F.	Excellent	Translucent to opaque, all colors.	Excellent deep draw, easily formed.
Polycarbonate	440 to 475°F.	Good	Good clarity and colorability.	Close temperature control required, excellent impact strength.
Acetal	365 to 390°F.	Fair	All translucent and opaque colors.	Very close control of heating temperatures required. Fairly difficult to process.
Styrene Acrylonitrile	430 to 450°F.	Good	Good clarity, wide range of colors.	Fairly brittle but forms easily.
Ethyl Cellulose	225 to 290°F.	Excellent	Brilliant clarity, transparent to opaque colors.	Easily formed, provides good deep drawing.

SPECIFIC GRAVITY OF PLASTICS

Specific gravity is the ratio of the weight of a material compared to the weight of an equal volume of water.

POLYMER	SPECIFIC GRAVITY
1. Acetal	1.42
2. Acrylic	1.18
3. Acrylonitrile Butadiene Styrene	1.04
4. Cellulose Acetate	1.30
5. Diallyl Phthalate	1.34–1.78
6. Epoxy (cast)	1.11
7. Ionomer	0.93–0.96
8. Melamine Formaldehyde	1.48
9. Nylon	1.13–1.20
10. Phenol Formaldehyde	1.25–1.30
11. Phenoxy	1.17
12. Polyallomer	0.89–0.90
13. Polycarbonate	1.20
14. Polyester	1.01–1.20
15. Polyethylene, L. D.	0.91–0.92
16. Polyethylene, H. D.	0.94–0.96
17. Polyphenylene Oxide	1.08
18. Polypropylene	0.89–0.90
19. Polystyrene	1.04–1.10
20. Polysulfone	1.24
21. Polyurethane	1.11–1.25
22. Polyvinyl Chloride	1.16–1.35
23. Silicone	1.75
24. Tetrafluoroethylene	2.14–2.20

COMPRESSIVE STRENGTH OF PLASTICS

The compression force required to rupture or deform a given specimen.

POLYMER	psi
1. Epoxy	35,000
2. Phenol Formaldehyde	34,000
3. Melamine Formaldehyde	30,000
4. Cellulose Acetate	28,000
5. Polyester	25,000
6. Diallyl Phthalate	23,000
7. Polyurethane	20,000
8. Acrylic	16,000
9. Polyphenylene Oxide	16,000
10. Polystyrene	15,000
11. Silicone	15,000
12. Polysulfone	13,900
13. Acetal	13,000
14. Nylon	13,000
15. Polycarbonate	12,500
16. Phenoxy	11,000
17. Polyvinyl Chloride	11,000
18. Acrylonitrile Butadiene Styrene	10,000
19. Ionomer	8,000
20. Polypropylene	7,000
21. Polyallomer	4,000
22. Polyethylene, H. D.	3,200
23. Polyethylene, L. D.	2,300
24. Tetrafluoroethylene	1,700

TENSILE STRENGTH OF PLASTICS

The force necessary to pull the specimen apart.

POLYMER	psi
1. Polyester (glass filled)	30,000
2. Silicone (asbestos filled)	28,000
3. Epoxy (glass filled)	17,000
4. Nylon	15,000
5. Polysulfone	10,200
6. Polyphenylene Oxide	9,600
7. Polycarbonate	9,500
8. Acrylic	9,000
9. Acetal	8,800
10. Phenol Formaldehyde	8,500
11. Phenoxy	8,500
12. Melamine Formaldehyde	8,300
13. Cellulose Acetate	8,000
14. Acrylonitrile Butadiene Styrene	7,000
15. Polystyrene	7,000
16. Diallyl Phthalate	6,000
17. Polyurethane	6,000
18. Polypropylene	5,300
19. Polyethylene, H. D.	5,000
20. Tetrafluoroethylene	5,000
21. Polyvinyl Chloride	4,800
22. Ionomer	4,000
23. Polyallomer	3,500
24. Polyethylene, L. D.	2,000

WATER ABSORPTION OF PLASTICS

1/8 in. thick sample over 24 hour period.

POLYMER	PERCENT ABSORBED
1. Tetrafluoroethylene	0.00
2. Polyallomer	0.01
3. Polyethylene, H. D.	0.01
4. Polypropylene	0.01
5. Polyethylene, L. D.	0.015
6. Polystyrene	0.04
7. Polyphenylene Oxide	0.06
8. Epoxy	0.10
9. Silicone	0.12
10. Phenoxy	0.13
11. Polycarbonate	0.15
12. Acetal	0.22
13. Polysulfone	0.22
14. Acrylic	0.30
15. Ionomer	0.30
16. Acrylonitrile Butadiene Styrene	0.35
17. Diallyl Phthalate	0.35
18. Melamine Formaldehyde	0.35
19. Polyvinyl Chloride	0.40
20. Polyester	0.55
21. Phenol Formaldehyde	0.60
22. Polyurethane	0.80
23. Nylon	1.50
24. Cellulose Acetate	3.80

IMPACT STRENGTH

The mechanical energy absorbed by a standard test sample during fracture by a blow from a pendulum hammer. Comparison values.

POLYMER	FOOT POUNDS
1. Polycarbonate	8.0
2. Polyvinyl Chloride	5.5
3. Polyurethane	5.0 to flexible
4. Acrylonitrile Butadiene Styrene	4.2
5. Phenol Formaldehyde	4.0
6. Ionomer	3.5
7. Polyallomer	3.2
8. Polyethylene, H. D.	3.0
9. Tetrafluoroethylene	3.0
10. Polyphenylene Oxide	2.4
11. Cellulose Acetate	2.2
12. Polypropylene	2.0
13. Polysulfone	1.9
14. Phenoxy	1.8
15. Nylon	1.5
16. Acetal	1.4
17. Epoxy	1.0
18. Polyester	0.8
19. Polystyrene	0.7
20. Diallyl Phthalate	0.5
21. Acrylic	0.33
22. Melamine Formaldehyde	0.30
23. Polyethylene, L. D.	flexible
24. Silicone	flexible

CONTINUOUS RESISTANCE TO HEAT IN DEGREES FAHRENHEIT WHILE IN USE

POLYMER	TEMPERATURE IN DEG. F.
1. Acetal	220
2. Acrylic	180
3. Acrylonitrile Butadiene Styrene	200
4. Cellulose Acetate	180
5. Diallyl Phthalate	425
6. Epoxy	400
7. Ionomer	220
8. Melamine Formaldehyde	210
9. Nylon	325
10. Phenol Formaldehyde	450
11. Phenoxy	170
12. Polyallomer	210
13. Polycarbonate	250
14. Polyester	250
15. Polyethylene, L. D.	200
16. Polyethylene, H. D.	250
17. Polyphenylene Oxide	375
18. Polypropylene	300
19. Polystyrene	170
20. Polysulfone	325
21. Polyurethane	200
22. Polyvinyl Chloride	175
23. Silicone	600
24. Tetrafluoroethylene	550

NATURAL COLOR OF PLASTIC MOLDING COMPOUNDS

POLYMER	COLOR
1. Acetal	Light Tan Opaque
2. Acrylic	Crystal Clear
3. Acrylonitrile Butadiene Styrene	Light Tan Opaque
4. Cellulose Acetate	Clear
5. Diallyl Phthalate	Gray Opaque
6. Epoxy	Clear
7. Ionomer	Clear
8. Melamine Formaldehyde	Gray Opaque
9. Nylon	Off White Opaque
10. Phenol Formaldehyde	Dark Gray Opaque
11. Phenoxy	Crystal Clear
12. Polyallomer	Milky White
13. Polycarbonate	Clear
14. Polyester	Clear
15. Polyethylene, L. D.	Milky White
16. Polyethylene, H. D.	Milky White
17. Polyphenylene Oxide	Opaque Beige
18. Polypropylene	Milky White
19. Polystyrene	Crystal Clear
20. Polysulfone	Amber Clear
21. Polyurethane	Opaque Amber
22. Polyvinyl Chloride	Light Bluish Clear
23. Silicone	Variable
24. Tetrafluoroethylene	White Opaque

AVERAGE COMPRESSION MOLDING TEMPERATURES

POLYMER	UNITS IN DEG. F.
1. Acetal	350
2. Acrylic	325
3. Acrylonitrile Butadiene Styrene	350
4. Cellulose Acetate	280
5. Diallyl Phthalate	300
6. Epoxy	300
7. Ionomer	325
8. Melamine Formaldehyde	325
9. Nylon	500
10. Phenol Formaldehyde	325
11. Phenoxy	350
12. Polyallomer	300
13. Polycarbonate	450
14. Polyester	300
15. Polyethylene, L. D.	300
16. Polyethylene, H. D.	300
17. Polyphenylene Oxide	425
18. Polypropylene	350
19. Polystyrene	300
20. Polysulfone	550
21. Polyurethane	425
22. Polyvinyl Chloride	350
23. Silicone	325
24. Styrene Acrylonitrile	325

COMMON SOLVENT CEMENTS FOR THERMOPLASTICS

RESIN	SOLVENT CEMENT
Acrylic	Methylene Chloride
	Ethylene Dichloride
ABS	Methyl Ethyl Ketone
* Cellulosics	Methyl Alcohol
	Methyl Ethyl Ketone
Ethyl Cellulose	Ethylene Dichloride
Polycarbonate	Methylene Chloride
Polyphenylene Oxide	Chloroform
	Toluene
	Ethylene Dichloride
Polystyrene	Methyl Ethyl Ketone
	Methylene Chloride
Polyvinyl Chloride	Tetrahydrofuran

(* Cellulose acetate and cellulose acetate butyrate)

DRILL SPEED AND POINT ANGLE FOR DRILLING PLASTICS WITH STANDARD TWIST DRILLS

POLYMER	SPEED (rpm)	POINT ANGLE IN DEGREES
Acetal	300–800	118
Acrylic	1500–2500	95
Acrylonitrile Butadiene Styrene	500–900	118
Cellulose Acetate	1000–2000	90
Diallyl Phthalate	600–2000	90
Epoxy (Filled)	500–1000	118
Ionomer	300–800	95
Melamine Formaldehyde	600–2000	90
Nylon	900–1500	70–90
Phenol Formaldehyde	600–2000	90
Polyallomer	1000–3000	90
Polycarbonate	300–800	118
Polyester	500–2000	90
Polyethylene	1000–3000	70–90
Polyphenylene Oxide	500–900	100
Polypropylene	1000–3000	70–90
Polystyrene	1500–2500	95
Polysulfone	1000–2500	95
Polyurethane (Rigid)	1000–3000	70–90
Polyvinyl Chloride	900–2000	118
Tetrafluoroethylene	1000–2000	90

CIRCULAR SAWING PLASTICS

POLYMER	TYPE OF BLADE	TEETH PER INCH	SPEED (FT. PER MIN.)
Acetal	Hollow Ground	4–6	8000
Acrylic	Hollow Ground	3–4	3000
Acrylonitrile Butadiene Styrene	Combination	4–6	4000
Cellulose Acetate	Combination	4–6	3000
Diallyl Phthalate	Carbide Tip	8–10	2000
Epoxy (Filled)	Carbide Tip	8–10	3000
Ionomer	Hollow Ground	6–8	6000
Melamine Formaldehyde	Carbide Tip	8–10	5000
Nylon	Hollow Ground	8–10	5000
Phenol Formaldehyde	Carbide Tip	8–10	3000
Polyallomer	Hollow Ground	4–6	9000
Polycarbonate	Hollow Ground	4–6	8000
Polyester	Carbide Tip	8–10	5000
Polyethylene	Hollow Ground	4–6	9000
Polyphenylene Oxide	Hollow Ground	6–8	5000
Polypropylene	Hollow Ground	8–10	9000
Polystyrene	Hollow Ground	4–6	2000
Polysulfone	Hollow Ground	3–4	3000
Polyurethane (Rigid)	Hollow Ground	3–4	4000
Polyvinyl Chloride	Hollow Ground	4–6	3000
Tetrafluoroethylene	Hollow Ground	4–6	8000

NOTE: Softer polymers, such as polyethylene, are more easily and accurately cut on a squaring shear.

BAND SAWING PLASTICS

POLYMER	TYPE OF BLADE	TEETH PER INCH	SPEED (FT. PER MIN.)
Acetal	Metal	10–16	1500
Acrylic	Metal	5–7	2000–4000
Acrylonitrile Butadiene Styrene	Wood–Skip Tooth	5–7	1000–3000
Cellulose Acetate	Wood–Skip Tooth	4–6	1000–1500
Diallyl Phthalate	Metal	10–18	2000
Epoxy (Filled)	Metal	10–18	1500
Ionomer	Wood–Skip Tooth	4–6	1000–1500
Melamine Formaldehyde	Metal	10–18	2000
Nylon	Wood–Skip Tooth	4–6	1000
Phenol Formaldehyde	Metal	10–18	1500
Polyallomer	Wood–Skip Tooth	4–6	1200–1500
Polycarbonate	Metal	10–18	1500
Polyester	Metal	10–18	4000
Polyethylene	Wood–Skip Tooth	4–6	1200–1500
Polyphenylene Oxide	Wood–Skip Tooth	6–9	2000–3000
Polypropylene	Wood–Skip Tooth	4–6	1200–1500
Polystyrene	Metal	10–18	3000–4000
Polysulfone	Metal	6–9	2000–3000
Polyurethane (Rigid)	Wood–Skip Tooth	4–6	1000–2000
Polyvinyl Chloride	Metal	6–9	2000–3000
Tetrafluoroethylene	Wood–Skip Tooth	5–7	1200–1500

MACHINING PROPERTIES OF MOLDED PLASTICS

POLYMER	RATING
1. Acetal	Excellent
2. Acrylic	Good
3. Acrylonitrile Butadiene Styrene	Good to Excellent
4. Cellulose Acetate	Excellent
5. Diallyl Phthalate	Fair
6. Epoxy	Fair
7. Ionomer	Good
8. Melamine Formaldehyde	Fair
9. Nylon	Excellent
10. Phenol Formaldehyde	Fair to Good
11. Phenoxy	Excellent
12. Polyallomer	Good
13. Polycarbonate	Excellent
14. Polyester	Good
15. Polyethylene, L. D.	Good
16. Polyethylene, H. D.	Excellent
17. Polyphenylene Oxide	Excellent
18. Polypropylene	Good
19. Polystyrene	Good
20. Polysulfone	Excellent
21. Polyurethane	Fair to Excellent
22. Polyvinyl Chloride	Excellent
23. Silicone	Fair
24. Tetrafluoroethylene	Excellent

TYPICAL FILLERS USED IN PLASTICS

POLYMER	FILLER
Acetal	Glass fibers, TFE fibers
Acrylonitrile Butadiene Styrene	Glass fibers
Diallyl Phthalate	Glass fibers
Epoxy	Clay, silica, glass fibers, metal powders
Melamine Formaldehyde	Flock, asbestos, fabric, glass fibers, cellulose fibers
Nylon	Glass fibers, asbestos
Phenol Formaldehyde	Wood flour, asbestos, mica, glass fibers, cotton flock, metal powders
Polycarbonate	Glass fibers
Polyester	Clay, glass fibers, woven cloth, asbestos
Polyethylene	Asbestos, metal powders
Polyphenylene Oxide	Glass fibers
Polypropylene	Asbestos, glass fibers
Polystyrene	Glass fibers
Polyvinyl Chloride	Asbestos, clay
Silicone	Asbestos, glass fiber, quartz
Urea Formaldehyde	Cellulose fibers, asbestos

CHARACTERISTICS OF PLASTIC FOAMS

POLYMERS	DENSITY LBS./CU.FT.	TYPE	CELL STRUCTURE	AVAILABLE FORM	CHARACTERISTIC USES
ABS	31.0–62.0	Thermoplastic, rigid.	Closed	Pellets	Picture frames, ice buckets, wall plaques.
Cellulose Acetate	6.0–8.0	Thermoplastic, rigid.	Closed	Boards and rods.	Life buoys, aircraft floats, fuel tank floats.
Epoxy	2.0–23.0	Thermoset, rigid.	Closed	Liquid and precast sheets and blocks.	Acoustical insulation, refrigerator doors, pontoons, gun stocks, furniture.
Ionomer	2.0–20.0	Thermoplastic, flexible.	Closed	Sheets and rods.	Marine floats, gaskets, insulation, packaging.
Phenolic	0.1–22.0	Thermoset, rigid.	Open and closed.	Liquid foam–in–place resin.	Cores for boat hulls, pipe insulation, cores for plywood, insulation.
Polyethylene	2.0–35.0	Thermoplastic, rigid.	Closed	Sheets, rods, tubing, molded parts.	Package cushioning, life jackets, boat bumpers, gaskets.
Polystyrene	1.0–10.0	Thermoplastic, rigid.	Closed	Expanded beads, boards and blocks.	Hot and cold drink cups, packaging, food containers, insulation.
Polyurethane	1.5–70.0	Thermoset, flexible or rigid.	Open and closed	Foam–in–place liquids, boards and blocks.	Freezer insulation, marine flotation, furniture cushioning, packaging.
Polyvinyl Chloride	3.0–45.0	Thermoplastic, flexible or rigid.	Open and closed	Sheets, molded shapes, and expandable beads.	Thermal insulation, athletic gear, ice buckets, flooring, carpet backing.
Silicone	9.6–31.0	Thermoset, rigid or flexible.	Open and closed.	Liquid and sheet.	Plastic surgery, heat sealing blankets, sponges.

HARDNESS OF PLASTICS MOLDING COMPOUNDS

A COMPARISON OF A PLASTICS RESISTANCE TO INDENTATION

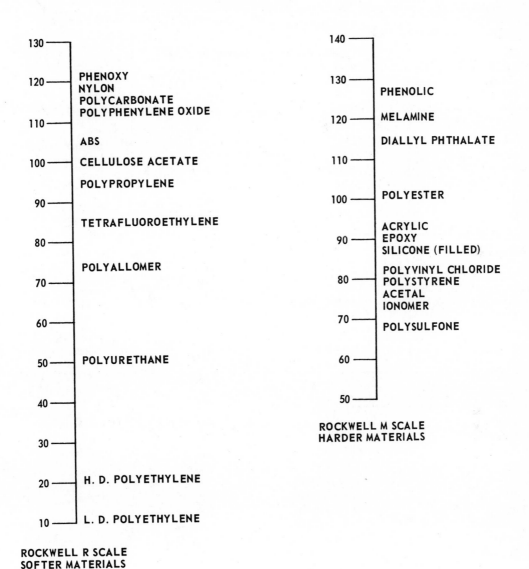

ROCKWELL R SCALE	ROCKWELL M SCALE
130 —	140 —
120 — PHENOXY / NYLON / POLYCARBONATE / POLYPHENYLENE OXIDE	130 — PHENOLIC
110 — / ABS	120 — MELAMINE / DIALLYL PHTHALATE
100 — CELLULOSE ACETATE / POLYPROPYLENE	110 —
90 — / TETRAFLUOROETHYLENE	100 — POLYESTER
80 — / POLYALLOMER	90 — ACRYLIC / EPOXY / SILICONE (FILLED)
70 —	80 — POLYVINYL CHLORIDE / POLYSTYRENE / ACETAL / IONOMER
60 —	70 — POLYSULFONE
50 — POLYURETHANE	60 —
40 —	50 —
30 —	
20 — H. D. POLYETHYLENE	
10 — L. D. POLYETHYLENE	

ROCKWELL R SCALE
SOFTER MATERIALS

ROCKWELL M SCALE
HARDER MATERIALS

289

Metric Tables

CONVERSION TABLE
ENGLISH TO METRIC

WHEN YOU KNOW: ⇩	MULTIPLY BY: * = Exact		TO FIND: ⇩
	VERY ACCURATE	APPROXIMATE	
LENGTH			
inches	* 25.4		millimetres
inches	* 2.54		centimetres
feet	* 0.3048		metres
feet	* 30.48		centimetres
yards	* 0.9144	0.9	metres
miles	* 1.609344	1.6	kilometres
WEIGHT			
grains	15.43236	15.4	grams
ounces	* 28.349523125	28.0	grams
ounces	* 0.028349523125	.028	kilograms
pounds	* 0.45359237	0.45	kilograms
short ton	* 0.90718474	0.9	tonnes
VOLUME			
teaspoon		5.0	millilitres
tablespoon		15.0	millilitres
fluid ounces	29.57353	30.0	millilitres
cups		0.24	litres
pints	* 0.473176473	0.47	litres
quarts	* 0.946352946	0.95	litres
gallons	* 3.785411784	3.8	litres
cubic inches	* 0.016387064	0.02	litres
cubic feet	* 0.028316846592	0.03	cubic metres
cubic yards	* 0.764554857984	0.76	cubic metres
AREA			
square inches	* 6.4516	6.5	square centimetre
square feet	* 0.09290304	0.09	square metres
square yards	* 0.83612736	0.8	square metres
square miles		2.6	square kilometre
acres	* 0.40468564224	0.4	hectares
TEMPERATURE			
Fahrenheit	*5/9 (after subtracting 32 from F. temp.)		Celsius

DICTIONARY OF TERMS

A

A-STAGE: An early stage in the reaction of a thermosetting resin in which the material is still linear in structure, soluble in certain liquids, and fuxible. (See also B- and C-STAGE.)

ABLATIVE PLASTICS: A material which absorbs heat (while part of it is being consumed by heat) through a thermal decomposition process known as pyrolysis, which takes place in the near surface layer exposed to heat.

ABSORPTION: (1) The penetration into the mass of one substance by another. (2) The process whereby energy is dissipated within a specimen placed in a field of radiation energy. Since processes other than absorption occur, for example, scattering, only a fraction of the energy removed from a beam is retained in the specimen.

ACCELERATOR: A substance that hastens a reaction. For example, sulfur-containing compounds which speed up the vulcanization of rubber. Also known as PROMOTOR. Often used to denote a substance that hastens the action of an initiator or catalyst.

ACCUMULATOR: A term used mainly with reference to blow molding equipment which designates an auxiliary ram extruder used to provide extremely fast parison delivery. The accumulator cylinder is filled with plasticated melt coming from the extruder between parison deliveries or "shots" and is stored or "accumulated" until the plunger is required to deliver the next parison.

ACETAL RESINS: Polymers containing the acetal linkage $\left(O - \overset{|}{\underset{|}{C}} - O \right)$. One example would be polyoxymethylene.

ACRYLIC ESTER: An ester of acrylic acid, or of a structural derivative of acrylic acid; for example, methyl methacrylate

$$\left(CH_2 = \overset{|}{\underset{CH_3}{C}} - \overset{\overset{O}{\|}}{C} - OCH_3 \right).$$

Reprinted by permission,
MODERN PLASTICS ENCYCLOPEDIA, McGraw-Hill, Inc.

ACRYLIC RESIN: A synthetic resin prepared from acrylic acid or from a derivative of acrylic acid.

ACRYLONITRILE: A monomer with the structure ($CH_2 = CHCN$). Its copolymer with butadiene is nitrile rubber, and several copolymers with styrene exist that are tougher than polystyrene. Its homopolymer is also used as a synthetic fiber.

ACRYLONITRILE - BUTADIENE - STYRENE (abbreviated ABS): Blends or copolymers of polystyrene or styrene-acrylonitrile copolymer with butadiene-acrylonitrile rubber.

ADIABATIC: An adjective used to describe a process or transformation in which no heat is added or allowed to escape from the system under consideration. It is used, somewhat incorrectly, to describe a mode of extrusion in which no external heat is added to the extruder, although heat may be removed by cooling to keep the output temperature of the melt passing through the extruder constant. The heat input in such a process is developed by the screw as its mechanical energy is converted to thermal energy.

ADDITION POLYMER: A polymer formed by a chain reaction, for example, polyethylene and polystyrene.

ADDITION POLYMERIZATION: Polymerization of monomers by a chain mechanism involving active centers on the growing chain. Frequently done with unsaturated monomers and usually called vinyl polymerization when unsaturated monomer contains $CH_2 = \overset{|}{\underset{|}{C}}$ group.

AGING: The change of a material with time under defined environmental conditions, leading to improvement or deterioration of properties.

AIR-ASSIST FORMING: A method of thermoforming in which airflow or air pressure is employed to partially preform the sheet immediately before final pulldown onto the mold using vacuum.

AIR GAP: In extrusion coating, the distance from die opening to nip formed by the pressure roll and the chill roll.

AIR RING: A circular manifold used to distribute an even flow of the cooling medium, air, onto a hollow tubular form passing through the center of the ring. In blown tubing, the air cools the tubing uniformly to provide uniform film thickness.

AIR-SLIP FORMING: A variation of snapback forming in which the male mold is enclosed in a box in such a way that when the mold moves forward toward the hot plastic, air is trapped between the mold and the plastic sheet. As the mold advances, the plastic is kept away from it by the air cushion formed as described above, until the full travel of the mold is reached, at which point a vacuum is applied, destroying the cushion and forming the part against the plug.

ALKYD RESIN: Polyesters made from dicarboxylic acids and diols, primarily used as coatings, modified with vegetable oil, fatty acids, etc.

ALLOY: Composite material made by blending polymers or copolymers with other polymers or elastomers under selected conditions, for example, styrene-acrylonitrile copolymer resins blended with butadiene-acrylonitrile rubbers. See POLYBLEND.

ALLYL RESIN: Synthetic resin formed by polymerization of chemical compounds containing $CH_2 = CH - CH_2-$. The principal commercial commercial allyl resin is a casting material that yields allyl carbonate polymer.

ALTERNATING COPOLYMER: Copolymer in which the molecules of each monomer alternate in the polymer chain. See also BLOCK and RANDOM COPOLYMERS.

AMINO: Indicates the presence of a - NH_2 group.

AMORPHOUS PHASE: Devoid of crystallinity, no definite order. At processing temperatures, a plastic is normally in an amorphous state.

ANGLE PRESS: A hydraulic molding press equipped with horizontal and vertical rams, and specially designed for the production of complex moldings containing deep undercuts.

ANIONIC POLYMERIZATION: Polymerization which is propagated by carbon atom intermediates which contain an unshared pair of electrons and are negatively charged (carbanions).

ANNEALING: A process of holding a material at an elevated temperature below its melting point to permit stress relaxation without distortion of shape. It is often used on molded articles to relieve stresses set up by flow into the mold.

ANTIOXIDANT: Substances which prevent or slow down oxidation of a polymeric material exposed to air.

ANTISTATIC AGENTS: Agents which minimize static electricity in plastics. Such agents are of two basic types: (1) metallic devices which come into contact with the plastics conducting the static to earth. Such devices give complete neutralization at the time, but because they do not modify the surface of the material it can become prone to further static during subsequent handling; (2) chemical additives which, when mixed with the compound during processing, give a reasonable degree of protection to the finished products.

ARC RESISTANCE: Time required for a given applied electrical voltage to render the surface of a material conductive because of carbonization by the arc discharge.

ATACTIC: Lack of structural regularity. Random placement of side chain substituents with respect to a vinyl polymer backbone.

AUTOACCELERATION: The increase in rate of polymerization and molecular weight of some vinly monomers polymerized in bulk or concentrated solution. Due to increase in viscosity of the reaction medium as the reaction proceeds. This impedes termination but does not appreciably affect propagation. Often called the Trommsdorff effect.

AUTOCLAVE: (1) Closed vessel for conducting chemical reactions under high pressure and temperature; (2) in low pressure laminating, a round or cylindrical container in which heat and pressure can be applied to resin-impregnated paper or fabric positioned in layers over a mold.

AUTOCLAVE MOLDING: Modification of the pressure bag method for molding reinforced plastics. After lay-up, entire assembly is placed in steam autoclave at 50 to 100 psi. Additional pressure achieves higher reinforcement loadings and improved removal of air.

AUTOMATIC MOLD: A mold for injection or compression molding that repeatedly goes through the entire cycle, including ejection, without human assistance.

AVERAGE MOLECULAR WEIGHT: Most synthetic polymers are a mixture of individual chains of many different sizes, hence, a molecular weight assigned to such a mixture is of necessity an average molecular weight.

B

B-STAGE: An intermediate stage in the reaction of a thermosetting resin in which the material softens when heated and swells in contact with certain liquids but does not entirely fuse or dissolve. Resins in thermosetting molding compounds are usually in this stage. See also A-STAGE and C-STAGE.

BACK PRESSURE: The viscosity resistance of a material to continued flow when a mold is closing. In extrusion, it is the resistance to the forward flow of molten material.

BACK-PRESSURE-RELIEF PORT: An opening from an extrusion die for the escape of any excess material.

BACK TAPER: Reverse draft used in mold to prevent molded article from drawing freely. See UNDERCUT.

BACKING PLATE: In injection molding, a plate used as a support for the cavity blocks, guide pins, bushings, etc.

BAFFLE: A device used to restrict or divert the passage of fluid through a pipe line or channel. In hydraulic systems, the device, which often consists of a disc with a small central perforation, restricts the flow of hydraulic fluid in a high-pressure line. A common location for the disc is in a joint in the line. When applied to molds, the term describes a plug or similar device which is located in a steam or water channel in the mold and designed to divert and restrict the flow to a desired path.

BAG MOLDING: A method of applying pressure during bonding or molding, in which a flexible cover, usually in connection with a rigid die or mold, exerts pressure on the material being molded, through the application of air pressure or drawing of a vacuum.

BAKELITE: A Union Carbide trade name to denote a thermoset synthesized by the condensation of phenol with formaldehyde. See PHENOLIC RESINS.

BALANCED RUNNER: A system designed to place all cavities the same distance from the sprue.

BANBURY: An internal type mixer for compounding materials composed of a pair of rotors which grind the materials to form a homogeneous blend.

BLANKING: The cutting of flat sheet stock to shape by striking it sharply with a punch while it is supported on a mating die. Punch presses are used. Also called DIE CUTTING.

BLEED: To give up color when in contact with water or a solvent; undesired movement of certain materials in a plastic (for example, plasticizers in vinyl) to the surface of the finished article or into an adjacement material.

BLISTER: A raised area on the surface of a molded plastic caused by the pressure of gases inside it on its incompletely hardened surface.

BLOCK COPOLYMER: Copolymer (- AAAA - BBBB -) in which the backbone consists of regions or blocks of one monomer (-AAAA-) along with regions or blocks of another monomer (-BBBB-).

BLOCKING: An undesired adhesion between touching layers of a material, such as occurs under moderate pressure during storage or use.

BLOW MOLDING: A method of fabrication in which a parison (hollow tube) is forced into the shape of the mold cavity by internal air pressure.

BLOW PRESSURE: The air pressure used to form a hollow part by blow molding.

BLOW RATE: The speed at which air enters the parison during the blow molding cycle.

BLOWING AGENTS: See FOAMING AGENTS.

BLOWUP RATIO: In blow molding, the ratio of the mold cavity diameter to the parison diameter. In blown tubing (film), the ratio of the final tube diameter (before gusseting, if any) to the original die diameter.

BLOWN TUBING: A thermoplastic film which is produced by extruding a tube, applying a slight internal pressure to the tube to expand it while still molten and subsequent cooling to set the tube. The tube is then flattened through guides and wound up flat on rolls. The size of blown tubing is determined by the flat width in inches as wound rather than by the diameter as would be the case with rigid types of tubing.

BLUEING: A mold blemish in the form of a blue oxide film which occurs on the polished surface of a mold as a result of the use of abnormally high mold temperatures.

BOSS: Protrusion on a plastic part designed to add strength, to facilitate alignment during assembly, to provide for fastenings, etc.

BOTTOM BLOW: A specific type of blow molding machine which forms hollow articles by injecting the blowing air into the parison from the bottom of the mold.

BOTTOM PLATE: Part of the mold which contains the heel radius and the push-up.

BRANCHED: In molecular structure of polymers, refers to side chains attached to the main chain. Side chains are generally short.

BREAKER PLATE: A perforated plate located at the rear end of an extruder head. It often supports the screens that prevent foreign particles from entering the die.

BREATHING: The opening and closing of a mold to allow gases to escape early in the molding cycle. Also called degassing. When referring to plastic sheeting, breathing indicates permeability to air, bubbles, voids or trapped globules of gas within a plastic part.

BUBBLER MOLD COOLING (INJECTION MOLDING): A method of cooling an injection mold in which a stream of cooling liquid flows continuously into a cooling cavity equipped with a coolant outlet which is normally positioned at the end opposite the inlet. Uniform cooling can be achieved in this manner.

BULK DENSITY: The mass per unit volume of a molding powder as determined in a reasonably large volume.

BULK FACTOR: Ratio of the volume of loose molding powder to the volume of the same weight of resin after molding.

BURNING RATE: A term describing the tendency of plastics to burn at given temperatures. Certain plastics, such as those based on shellac, burn readily at comparatively low temperatures. Others will melt or disintegrate without actually burning, or will burn only if exposed to direct flame. These latter are referred to as self-extinguishing.

BUSHING (EXTRUSION): The outer ring of a circular tubing or pipe die which forms the outer surface of the tube or pipe.

BUTADIENE: A diene monomer with the structure ($CH_2 = CH - CH = CH_2$). May be copolymerized with styrene and with acrylonitrile. Its homopolymer is used as a synthetic rubber.

BUTT-FUSION: A method of joining pipe, sheet, or other similar forms of a thermoplastic resin where the ends of the two pieces to be joined are heated to the molten state and then rapidly pressed together to form a homogenous bond.

C

C-STAGE: The final stage in the reactions of a thermosetting resin in which the material is insoluble and infusible. Thermosetting resins in a fully cured plastic are in this stage. See A-STAGE and B-STAGE.

CALENDER: (v) To prepare sheets of material by pressure between two or more counterrotating rolls. (n) The machine performing this operation.

CARBON BLACK: A black pigment produced by the incomplete burning of natural gas or oil. It is widely used as a filler or pigment, particularly in the rubber industry. It possesses useful untraviolet protective properties.

CASEIN: A protein material precipitated from skimmed milk by the action of dilute acid. Acid casein is a raw material used in the manufacture of buttons and buckles.

CAST: (1) To form a "plastic" object by pouring a fluid monomer-polymer solution into an open mold where it finishes polymerizing; (2) forming plastic film and sheet by pouring the liquid resin onto a moving belt or by precipitation in a chemical bath.

CASTING: (n) The finished product of a casting operation; should not be used for molding.

CASTING AREA: The moldable area of a thermoplastic in square inches for a given thickness and under a given set of injection molding conditions. Casting area is a measure of flow under actual molding conditions where flow is unrestricted by cavity boundaries.

CATALYST: A substance which speeds up the polymerization or cure of a compound when added in minor quantity as compared to the amounts of primary reactants, providing it does not become a component part of the chain; otherwise it is referred to as an initiator. See HARDENER, INHIBITOR, PROMOTOR.

CAVITY: Depression in a mold made by casting, machining, hobbing, or a combination of these methods; depending on number of such depressions, molds are designated as single-cavity or multicavity.

CELLULAR PLASTICS: See CHEMICALLY FOAMED PLASTIC.

CELLULOID: A thermoplastic material made from plasticized cellulose nitrate and camphor. Alcohol is normally employed as a volatile solvent in order to assist plasticization, and is subsequently removed.

CELLULOSE: A naturally occurring polysaccharide made up solely of glucose units and found in most plants; the main constituent of dried woods, jute, flax, hemp, ramie, etc. Cotton is almost pure cellulose; however, in many other important natural cellulosic materials, the cellulose is associated with sizable quantities of impurities including lignin (a natural resin) and various hexosans, pentosans, and polyuronides collectively called hemicelluloses.

CELLULOSE ACETATE: An acetic acid ester of cellulose. It is obtained by the action, under rigidly controlled conditions, of acetic acid and acetic anhydride on purified cellulose usually obtained from cotton linters. All three available hydroxyl groups in each glucose unit of the cellulose can be acetylated, but in the preparation of cellulose acetate it is usual to acetylate fully and then to lower the acetyl value (expressed as acetic acid) to 52-56 percent by partial hydrolysis. When compounded with suitable plasticizers, it gives a tough thermoplastic material.

CELLULOSE ACETATE BUTYRATE: An ester of cellulose made by the action of a mixture of acetic and butyric acids and their anhydrides on purified cellulose. It is used in the manufacture of plastics which are similar in general properties to cellulose acetate but are tougher and have better moisture resistance and dimensional stability.

CELLULOSE ESTER: A derivative of cellulose in which the free hydroxyl groups attached to the cellulose chain have been replaced wholly or in part by acidic groups, for example, nitrate or carboxylate groups. Esterification is effected by the use of a mixture of an acid with its anhydride in the presence of a catalyst, such as sulfuric acid. Mixed esters of cellulose, like cellulose acetate butyrate, are prepared by the use of mixed acids and mixed anhydrides. Esters and mixed esters, a wide range of which are known, differ in their compatibility with plasticizers, in molding properties, and in physical characteristics. These esters and mixed esters are used in making thermoplastic molding compositions.

CELLULOSE NITRATE (NITROCELLULOSE): A nitric acid ester of cellulose manufactured by the action of a mixture of sulfuric acid and nitric acid on cellulose, such as purified cotton linters.

CELLULOSE PROPIONATE: An ester of cellulose made by the action of propionic acid and its anhydride on purified cellulose. It is used as the basis of a thermoplastic molding material.

CELLULOSE TRIACETATE: A cellulosic material made by reacting purified cellulose with acetic anhydride in the presence of a catalyst. All three hydroxyl groups in each glucose unit of the cellulose are esterfied. It is used in the form of films and fibers. Films and sheets are cast from clear solutions onto "drums" with highly polished surfaces. The film, which is of excellent clarity, has high tensile strength, good heat resistance, and dimensional stability.

CENTER GATED MOLD: An injection mold wherein the cavity is filled with resin through an orifice interconnecting the nozzle and the center of the cavity area. Normally, this orifice is located at the botton of the cavity when forming items such as containers, tumblers, bowls, etc.

CENTIPOISE: A unit of viscosity, conviently and approximately defined as the viscosity of water at room temperature. The following table of approximate viscosities at room temperature may be useful for rough comparisons:

LIQUID	VISCOSITY IN CENTIPOISES
water	1
kerosene	10
motor oil SAE 10	100
caster oil; flycerin	1,000
corn syrup	10,000
molasses	100,000

CENTRIFUGAL CASTING: A method of forming thermoplastic resins in which the granular resin is placed in a rotatable container, heated to a molten condition by the transfer of heat through the walls of the container, and rotated so that the centrifugal force induced will force the molten resin to conform to the configuration of the interior surface of the container. Used to fabricate large diameter pipes and similar cylindrical items.

CHAIN LENGTH: See DEGREE OF POLYMERIZATION.

CHARGE: The measurement or weight of material used to load a mold at one time or during one cycle.

CHASE: An enclosure of any shape, used to: (a) shrink-fit parts of a mold cavity in place; (b) prevent spreading or distortion in hobbing; (c) enclose an assembly or two or more parts of a split cavity block.

CHEMICALLY FOAMED PLASTIC: A cellular plastic whose structure is produced by gases generated from the chemical interaction of its constituents, for example, polyurethanes.

CHILL ROLL: A cored roll, usually temperature controlled with circulating water, which cools the web before winding. For chill roll (cast) film, the surface of the roll is highly polished. In extrusion coating, either a polished or a matte surface may be used depending on the surface desired on the finished coating.

CHILL ROLL EXTRUSION (OR CAST FILM EXTRUSION): The extruded film is cooled while being drawn around two or more highly polished chill rolls cored for water cooling for exact temperature control.

CHOKED NECK: Narrowed or constricted opening in the neck of a container.

CLAMPING PLATE: A plate fitted to a mold and used to fasten the mold to a machine.

CLAMPING PRESSURE: In injection molding and in transfer molding, the pressure which is applied to the mold to keep it closed, in opposition to the fluid pressure of the compressed molding material.

CLOSED-CELL FOAM: A cellular plastic in which there is a predominance of noninterconnecting cells.

CLEARANCE: A controlled distance by which one part of an object is kept separated from another part.

COATING: See specific type of coating such as CURTAIN, EXTRUSION, KISS-ROLL, SPRAY.

COATHANGER DIE: One basic type of extrusion slot die shaped internally like a coat hanger to gain better distribution across the full width of a sheet extrusion.

COATING WEIGHT: The weight of coating per unit area. In the United States usually "per ream," which is 500 sheets 24 x 36 in. (3000 sq. ft.), but sometimes 1000 sq. ft.

COLD FLOW: See CREEP.

COLD MOLDING: A procedure in which a composition is shaped at room temperature and cured by subsequent baking.

COLD SLUG: The first material to enter an injection mold; so called because in passing through the sprue orifice it is cooled below the effective molding temperature.

COLD SLUG WELL: Space provided directly opposite the sprue opening in an injection mold to trap the cold slug.

COLD STRETCH: Pulling operation with little or no heat, usually on extruded filaments, to improve tensile properties.

COLLAPSE: Contraction of the walls of a container, for example upon cooling, leading to a permanent indentation.

COMPRESSION MOLD: A mold which is open when the material is introduced and which shapes the material by heat and by the pressure of closing.

COMPRESSION MOLDING: A technique of thermoset molding in which the molding compound (generally preheated) is placed in the open mold cavity, mold is closed, and heat and pressure (in the form of a downward moving ram) are applied until the material has cured.

COMPRESSION RATIO: In an extruder screw, the ratio of volume available in the first flight at the hopper to the last flight at the end of the screw.

COMPRESSIVE STRENGTH: Pressure load at failure of a shaped specimen divided by a cross-sectional area of the specimen, usually the original sectional area.

CONDENSATION: A chemical reaction in which two or more molecules combine with the separation of water or some other simple substance. If difunctional or higher functionality molecules react, the condensation process is called polycondensation. See also POLYMERIZATION.

CONDENSATION POLYMER: A polymer formed by polycondensation, for example, the alkyd, phenol formaldehyde, and urea formaldehyde resins.

CONDITIONING: The subjection of a material to a stipulated treatment so that it will respond in a uniform way to subsequent testing or processing. The term is frequently used to refer to the treatment given to specimens before testing.

CONSISTENCY: The resistance of a material to flow or permanent deformation when shearing stresses are applied to it. The term is generally used with materials whose deformations are not proportional to applied stresses. Viscosity is generally considered to be a similar internal friction that results in flow in proportion to the stress applied. See VISCOSITY.

CONTACT PRESSURE RESINS: Liquid resins which thicken or resinify on heating and, when used for bonding laminates, require little or no pressure.

CONTINUOUS TUBE PROCESS: A blow molding process using a continuous extrusion of tubing to feed into the blow molds as they clamp in sequence.

CONVERGENT DIE: A die in which the internal channels leading to the orifice are converging (only applicable to dies for hollow bodies).

COOLING CHANNELS: Channels or passageways located within the body of a mold through which a cooling medium can be circulated to control temperature.

COOLING FIXTURE: Block of metal or wood holding the shape of a molded piece which is used to maintain the proper shape or dimensional accuracy of a molding after it is removed from the mold until it is cool enough to retain its shape without further appreciable distortion. Also known as Shrink Fixture.

COPOLYMER: A polymetric system which contains two or more monomeric units. See also ALTERNATING, RANDOM, and BLOCK COPOLYMER.

CORE AND SEPARATOR: The center section of an extrusion die.

CORE: (1) The central member of a sandwich construction (can be honeycomb material, foamed plastic, or solid sheet) to which the faces of the sandwich are attached; the central member of a plywood assembly. (2) A channel in a mold for circulation of heat-transfer media. (3) Part of a complex mold that molds undercut parts. Cores are usually withdrawn to one side before the main sections of the mold open. Also called Core Pin.

CORE DRILL: A device for making cooling channels in a mold.

CRAZING: Fine cracks which may extend in a network on or under the surface or through a layer of a plastic material. Usually occurs in the presence of an organic liquid or vapor, with or without the application of mechanical stress.

CREEP: The dimensional change with time of a material under load, following the initial instantaneous elastic deformation. Creep at room temperature is called cold flow.

CROSSHEAD (EXTRUSION): A device generally employed in wire coating which is attached to the discharge end of the extruder cylinder, designed to facilitate extruding material at an angle. Normally, this is a 90 deg. angle to the longitudinal axis of the screw.

CROSS-LINKING: The formation of primary valence bonds between polymer molecules. When extensive, as in thermosetting resins, cross-linking makes one infusible, insoluble, super-molecule of all the chains.

CRYSTALLINITY: A state of molecular structure in some resins which denotes stereoregularity and compactness of the molecular chains forming the polymer. Normally can be attributed to the formation of solid crystals having a definite geometric form.

CULL: Material remaining in a transfer chamber after mold has been filled. Unless there is a slight excess in the charge, the operator cannot be sure cavity is filled. Charge is generally regulated to control thickness of cull.

CURE: The changing of the physical properties of a material by chemical reactions such as polycondensation, addition polymerization, or vulcanization; usually accomplished by the action of heat and catalysts, alone or in combination, with or without pressure.

CURING TEMPERATURE: Temperature at which a cast, molded, or extruded product, resin-impregnated reinforcement, adhesive, etc., is subjected to curing.

CURING TIME (MOLDING TIME): In the molding of thermosetting plastics, the interval of time between the instant of cessation of relative movement between the moving parts of a mold and the instant that pressure is released.

CURLING: A condition in which the parison curls upwards and outwards, sticking to the outer face of the die ring. Balance of temperatures between die and mandrel will normally relieve this problem.

CURTAIN COATING: A method of coating which may be employed with low viscosity resins or solutions, suspensions, or emulsions of resins in which the substrate to be coated is passed through and perpendicular to a freely falling liquid "curtain" (or "waterfall"). The flow rate of the falling liquid and linear speed the substrate passing through the curtain are coordinated in accordance with the thickness of coating desired.

CURVATURE: A condition in which the parison is not straight, but somewhat bending and shifting to one side, leading to a deviation from the vertical direction of extrusion. Centering of ring and mandrel can often relieve this defect.

CUT-OFF: The line where the two halves of a compression mold come together; also called Flash Groove or Pinch-off.

CYCLE: The complete, repeating sequence of operations in a process or part of a process. In molding, the cycle time is the period, or elapsed time, between a certain point in one cycle and the same point in the next cycle.

D

DAYLIGHT OPENING: Clearance between two press platens in the open position.

DEBOSSED: An indented or cut-in design or lettering on a surface.

DECKLE ROD: A small rod, or similar device, inserted at each end of the extrusion coating die which is used to adjust the length of the die opening.

DECORATIVE SHEET: A laminated plastics sheet used for decorative purposes in which the color and/or surface pattern is an integral part of the sheet.

DEFLASHING: Covers the range of finishing techniques used to remove the flash (excess material) on a plastic molding.

DEGASSING: See BREATHING.

DEGREE OF POLYMERIZATION (DP): The number of structural units or mers in the polymer molecule in a particular sample. The value is obtained from the molecular weight of the polymer divided by that of the mer. If "average" molecular weight is used, then the value is the "average" DP. In most polymers the DP must reach several thousand if worthwhile physical properties are to be had.

DELAMINATION: The separation of the layers in a laminate caused by the failure of the adhesive, or resin binder.

DELIQUESCENT: Capable of attracting moisture from the air.

DENSITY: Weight per unit volume of a substance, expressed in grams per cubic centimeter, pounds per cubic foot, etc.

DESTATICIZATION: Treating plastic materials to minimize their accumulation of static electricity and, consequently, the amount of dust picked up by the plastics because of such charges.

DIAPHRAGM GATE: Gate used in molding annular or tubular articles.

DIE-ADAPTOR: That part of an extrusion die which holds the die block.

DIE BLADES: Deformable members attached to a die body which determine the slot opening and which are adjusted to produce uniform thickness across the film or sheet produced.

DIE BLOCK: That part of an extrusion die which holds the forming bushing and core.

DIE BODY: That part of an extrusion die used to separate and form material.

DIE CUTTING: (1) Blanking; (2) Cutting shapes from sheet stock by striking it sharply with a shaped knife edge known as a "steel-rule die." Clicking and Dinking are other names for die cutting of this kind.

DIE GAP: The distance between the metal faces forming the die opening.

DIE LINES: Vertical marks on the parison caused by damage of die parts or contamination.

DIE SWELL RATIO: The ratio of the outer parison diameter (or parison thickness) to the outer diameter of the die (or die gap). Die swell ratio is influenced by polymer type, head construction, land length, extrusion speed, and temperature.

DIELECTRIC: Insulating material. In radio frequency preheating, dielectric may refer specifically to the material which is being heated.

DIELECTRIC CONSTANT: Normally the relative dielectric constant; for practical purposes, the ratio of the capacitance of an assembly of two electrodes separated solely by a dielectric material to its capacitance when the electrodes are separated solely by air.

DIELECTRIC HEATING (ELECTRONIC HEATING): The plastic to be heated forms the dielectric of a condenser to which is applied a high-frequency (20 to 80 mc) voltage. Process is used for sealing vinyl films and preheating thermoset molding compounds.

DIELECTRIC STRENGTH: The electric voltage gradient at which an insulating material is broken down or "arced through." In volts per mil of thickness.

DIMER: A substance comprised of two molecules of a monomer. The degree of polymerization is two.

DIP COATING: Applying a plastic coating by dipping the article to be coated into a tank of melted resin or plastisol, then chilling the adhering melt.

DISPERSANT: In an organosol, a liquid component which has a solvating or peptizing action on the resin, so as to aid in dispersing and suspending it.

DISPERSION: Finely divided particles of a material in suspension in another substance.

DIVERGENT DIE: A die in which the internal channels leading to the orifice are diverging (applicable only to dies for hollow bodies).

DOCTOR ROLL, DOCTOR BAR, DOCTOR BLADE: A device for regulating the amount of liquid material on the rollers of a spreader.

DOUBLE-SHOT MOLDING: A means of turning out two-color parts in thermoplastics materials by successive molding operations.

DOWEL: Pin used to maintain alignment between two or more parts of a mold.

DRAFT: The degree of taper of a side wall or the angle of clearance designed to facilitate removal of parts from a mold.

DRAPE ASSIST FRAME: In sheet thermoforming, a frame (made of anything from thin wires to thick bars) shaped to the peripheries of depressed areas of the mold and suspended above the sheet to be formed. During forming, the assist frame drops down, drawing the sheet tightly into the mold and thereby preventing webbing between high areas of mold and permitting closer spacing in multiple molds.

DRAPE FORMING: Method of forming thermoplastic sheet in which the sheet is clamped into a movable frame, heated, and draped over high points of a male mold. Vacuum is then pulled to complete the forming operation.

DRAW DOWN RATIO: The ratio of the thickness of the die opening to the final thickness of the product.

DRAWING: The process of stretching a thermoplastic to reduce its cross-sectional area, placing the polymer chains in a more orderly arrangement with respect to each other.

DRY COLORING: Method commonly used by fabricators for coloring plastics by tumble blending uncolored particles of the plastic material with selected dyes and pigments.

DRY SPOT: Area of incomplete surface film on laminated plastics; in laminated glass, an area over which the interlayer and the glass are not bonded.

DRY STRENGTH: The strength of an adhesive joint determined immediately after drying under specified conditions or after a period of conditioning in the standard laboratory atmosphere.

DUCTILITY: The extent to which a solid material can be drawn into a thinner cross section.

DUPLICATE CAVITY-PLATE: Removable plate that retains cavities, used where two-plate operation is necessary for loading inserts.

DUROMETER HARDNESS: Hardness of a material measured by Shore Durometer.

DWELL: A pause in the application of pressure to a mold, made just before the mold is completely closed, to allow the escape of gas from the molding material.

DYES: Synthetic or natural organic chemicals that are soluble in most common solvents. Characterized by good transparency, high tinctorial strength, and low specific gravity.

E

EJECTOR PIN (ON SLEEVE): A pin or thin plate that is driven into a mold cavity from the rear as the mold opens, forcing out the finished piece. Also Knock-out Pin.

EJECTOR PIN RETAINER PLATE: Retainer into which ejector pins are assembled.

EJECTOR PLATE: A plate which backs up the ejector pins and holds the ejector assembly together.

ELASTIC DEFORMATION: The part of the deformation of an object under load which is recoverable when the load is removed.

ELASTICITY: That property of a material by virtue of which it tends to recover its original size and shape after deformation. If the strain is proportional to the applied stress, the material is said to exhibit Hookean or ideal elasticity.

ELASTOMER: A material which at room temperature stretches under low stress to at least twice its length and snaps back to the original length upon release of stress.

ELECTROFORMED MOLD: A mold made by electroplating metal on the reverse pattern on the cavity. Molten steel may be then sprayed on the back of the mold to increase its strength.

ELONGATION: The fractional increase in length of a material stressed in tension.

EMBOSSING: Techniques used to create depressions of a specific pattern in plastic film and sheeting.

EMULSION: A suspension of extremely fine droplets of one liquid in another liquid.

ENCAPSULATING: Encasing of an article (usually an electronic component or the like) in a closed envelope of plastic, by immersing the object in a casting resin and allowing the resin to polymerize or, if hot, to cool.

ENGINEERING PLASTICS (colloq.): Plastics that lend themselves to engineering design, for example, gears, structural members, etc.

ENGRAVED-ROLL (OR GRAVURE) COATING: The amount of coating applied to the web is metered by the depth of the overall engraved pattern in a print roll. This process is frequently modified by interposing a resilient offset roll between the engraved roll and the web.

ENTRANCE ANGLE: Maximum angle at which the molten material enters the land area of the die, measured from the center line of the mandrel.

ENVIRONMENTAL STRESS CRACKING (ESC): The susceptibility of a thermoplastic to crack or craze under influence of chemical treatment and/or mechanical stress.

EPOXY RESINS: Based on ethylene oxide, its derivatives or homologs, epoxy resins form straight-chain thermoplastics and thermosetting resins. The condensation of bisphenol and epichlorohydrin yields, a thermoplastic which is converted to a thermoset by active hydrogen-containing compounds, such as polyamines, dianhydrides.

ESTER: The reaction product of an alcohol and an acid.

ETHYLENE-VINYL ACETATE: Copolymer of ethylene and vinyl acetate having many of the properties of polyethylene, but of considerably increased flexibility for its density. Elongation and impact resistance are also increased.

EXOTHERM: (1) The temperature/time curve of a chemical reaction giving off heat, particularly the polymerization of casting resins. (2) The amount of heat given off. The term has not been standardized regarding sample size, ambient temperature, degree of mixing, etc.

EXPANDED PLASTICS: See CHEMICALLY FOAMED PLASTICS.

EXTENDER: A substance added to a plastic composition to reduce the amount of the primary resin required per unit area.

EXTRUDATE: The product or material delivered by an extruder, such as film, pipe, the coating on wire, etc.

EXTRUSION: The compacting of a plastic material and the forcing of it through an orifice in more or less continuous fashion.

EXTRUSION COATING: The coating of a resin on a substrate by extruding a thin film of molten resin and pressing it onto or into the substrate without an adhesive.

F

FABRICATE: To work a material into a finished form by machining, forming, or other operation or to make flexible film or sheeting into end products by sewing, cutting, sealing or other operation.

FADEOMETER: An apparatus for determining the resistance of resins and other materials to fading. This apparatus accelerates the fading by subjecting the article to high intensity ultraviolet rays of approximately the same wavelength as those found in sunlight.

FALSE NECK: A neck construction which is additional to the neck finish of a container and which is only intended to facilitate the blow molding operation. Afterwards the false neck part is removed from the container.

FAMILY MOLD (INJECTION): A multicavity mold where each of the cavities forms one of the component parts of the assembled finished object.

FIBER: This term usually refers to relatively short lengths of very small cross sections of various materials. Fibers can be made by chopping filaments (converting). Staple fibers may be 1/2 to a few inches in length and usually 1 to 5 denier.

FILAMENT: A variety of fiber characterized by extreme length, which permits its use in yarn with little or no twist and usually without the spinning operations required for fibers.

FILL-AND-WIPE: Parts are molded with depressed designs; after application of paint, surplus is wiped off, leaving paint remaining only in depressed areas.

FILLER: A cheap, inert substance added to a plastic to make it less costly. Fillers may also improve physical properties, particularly hardness, stiffness, and impact strength. The particles are usually small, in contrast to those of reinforcement; but there is some overlap between the functions of the two types of material.

FILLET: A rounded filling of the internal angle between two molding surfaces.

FILM: An optional term for sheeting having a nominal thickness not greater than 0.010 in.

FIN: The web of material remaining in holes or openings in a molded part which must be removed in finishing.

FINISH: The plastic forming the opening of a container shaped to accommodate a specific closure. Also, the ultimate surface structure of an article.

FISH EYE: A fault in transparent or translucent plastics materials, such as film or sheet, appearing as a small globular mass and caused by incomplete blending of the mass with surrounding material.

FLAKE: Used to denote the dry, unplasticized base of cellulosic plastics.

FLAME RETARDANT RESIN: A resin which is compounded with certain chemicals to reduce or eliminate its tendency to burn. For polyethylene and similar resins, chemicals such as antimony trioxide and chlorinated paraffins are useful.

FLAME SPRAYING: Method of applying a plastic coating in which finely powdered fragments of the plastic, together with suitable fluxes, are projected through a cone of flame onto a surface.

FLAMMABILITY: Measure of the extent to which a material will support combustion.

FLASH: Extra plastic attached to a molding along the parting line which must be removed before the part can be considered finished.

FLASH LINE: A raised line appearing on the surface of a molding and formed at the junction of mold faces.

FLASH MOLD: A mold designed to permit excess molding material to escape during closing.

FLEXIBLE MOLDS: Molds made of rubber or elastomers used for casting plastics. They can be stretched to remove cured pieces with undercuts.

FLEXURAL STRENGTH: The strength of a material in bending, expressed as the tensile stress of the outermost fibers of a bent test sample at the instant of failure. With plastics, this value is usually higher than the straight tensile strength.

FLOCK: Short fibers of cotton, etc., used as fillers for molding materials.

FLOW: A qualitative description of the fluidity of a plastic during molding.

FLOW LINE (WELD LINE): A mark on a molded piece made by the meeting of two flow fronts during molding.

FLOW MARKS: Wavy surface appearance of an object molded from thermoplastic resins caused by improper flow of the resin into the mold.

FLUIDIZED BED COATING: A method of applying a coating of a thermoplastic resin to an article in which the heated article is immersed in a dense-phase fluidized bed of powdered resin and thereafter heated in an oven to provide a smooth, pinhole-free coating.

FLUORESCENT PIGMENTS: Pigments which absorb radiation of a given frequency and then emit radiation of a different frequency. Leads to a glowing effect.

FLUOROPLASTICS (FORMERLY CALLED FLUOROCARBONS): The family of plastics including polytetrafluoroethylene(PTFE); polychlorotrifluoroethylene (PCTFE); polyvinylidene fluoride and fluorinated ethylene propylene (FEP). They are characterized by properties including good thermal and chemical resistance and nonadhesiveness, and possess a low dissipation factor and low dielectric constant. Depending upon which of the flurocarbons is used, they are available as molding materials, extrusion materials, dispersions, film or tape.

FOAMING AGENTS: Chemicals added to plastics and rubber that generate inert gases on heating, causing the resin to assume a cellular structure.

FOAM-IN-PLACE: Refers to the deposition of foams which requires that the foaming machine be brought to the work which is "in place" as opposed to bringing the work to the foaming machine.

FOIL DECORATING: Molding paper, textile, or plastic foils printed with compatible inks directly into a plastic part so that the foil is visible below the surface of the part as integral decoration.

FORMING: A process in which the shape of plastic pieces such as sheets, rods, or tubes is changed to a desired configuration. The use of the term forming in plastics technology does not include such operations as molding, casting, or extrusion, in which shapes or articles are made from molding materials or liquids.

FRICTION CALENDERING: A process where an elastomeric compound is forced into the interstices of woven or cord fabrics while passing through the calender rolls.

FRICTION WELDING: A method of welding thermoplastics materials where the heat necessary to soften the components is provided by friction.

FROTHING: Technique for applying urethane foam in which blowing agents or tiny air bubbles are introduced under pressure into the liquid mixture of foam ingredients.

FURANE RESINS: Dark colored, thermosetting resins which are primarily liquids ranging from low-viscosity polymers to thick, heavy syrups. Based on furfural or furfuryl alcohol.

FUSE: In plastisol molding, to heat the plastisol to the temperature at which it becomes a single homogeneous phase. In this sense, Cure is the same as Fuse.

G

GATE: In injection and transfer molding, the orifice through which the melt enters the cavity. Sometimes the gate has the same cross section as the runner leading to it; often, it is severely restricted.

GEL: That part of a three-dimensional step-reaction (cross-linked) polymer which is insoluble in all nondegrading solvents. The sol portion remains soluble and may be extracted from the gel.

GEL COAT: A thin, outer layer of resin, sometimes with pigment, applied to a reinforced plastics molding as a cosmetic.

GEL POINT: The stage at which a liquid begins to exhibit elastic properties and increased viscosity. This stage may be conveniently observed from the inflection point on a viscosity-time plot. See GEL.

GLASS BONDED MICA: A moldable thermoplastic having a glass binder and mica filler. Also known as Ceramoplastic.

GLASS TRANSITION: The change in an amorphous polymer or in amorphous regions of a partially crystalline polymer from (or to) a viscous or rubbery condition to (or from) a hard and relatively brittle one. This transition generally occurs over a relatively narrow temperature region and is similar to the solidification of a liquid to a glassy state; it is not a phase transition. Not only do hardness and brittleness undergo rapid changes in this temperature region, but other properties such as thermal expansibility and specific heat also change rapidly. This phenomenon has been called second-order transition, glass transition, rubber transition, and rubbery transition.

GLASS TRANSITION TEMPERATURE (Tg): The temperature region in which the glass transition occurs. See GLASS TRANSITION. The measured value of the glass transition temperature depends to some extent on the method of test.

GLITTER (OR FILTER OR SPANGLES): A group of special decorative materials consisting of flakes large enough so that each separate flake produces a plainly visible sparkle or reflection. They are incorporated directly into the plastic during compounding.

GRANULAR STRUCTURE: Nonuniform appearance of finished plastic material due to retention of, or incomplete fusion of, particles of composition, either within the mass or on the surface.

GRIT BLASTED: A surface treatment of a mold in which steel grit or sand materials are blown to the walls of the cavity to produce a roughened surface. Air escape from mold is improved and special appearance of molded article is often obtained by this method.

GUIDE PINS. Devices that maintain proper alignment of force plug and cavity as mold closes.

H

HAND MOLD: A mold taken out of the press after each shot for part removal.

HARDENER: A substance or mixture of substances added to a monomeric or polymeric composition, or an adhesive to promote or control the curing reaction. Also designates a substance added to control degree of hardness of a cured system. See also CATALYST.

HARDNESS: The resistance of a plastics material to compression and indentation. Among the most important methods of testing this property are Brinell hardness, Rockwell hardness and Shore hardness.

HAZE: The degree of cloudiness in a plastics material.

HEAD: End section of blow molding machine (in a general extruder) in which melt is transformed into a hollow parison.

HEAT-DISTORTION POINT: The temperature at which a standard test bar deflects 0.010 in. under a stated load of either 66 or 264 psi.

HEAT-FREQUENCY HEATING: The heating of materials by dielectric loss in a high-frequency electrostatic field. The material is exposed between electrodes, and by absorption of energy from the electrical field, is heated quickly and uniformly throughout.

HEAT-SEALING: A method of joining plastic films by simultaneous application of heat and pressure to areas in contact. Heat may be supplied conductively or dielectrically.

HIGH POLYMER: A large molecule which is usually but not always comprised of repeat units of the low molecular weight species. Arbitrarily designated as having a molecular weight greater than 10,000.

HIGH-PRESSURE LAMINATES: Laminates molded and cured at pressures not lower than 1000 psi and more commonly in the range of 1200 to 2000 psi.

HOB: A master model in hardened steel used to sink the shape of a mold into a soft steel block.

HOMOPOLYMER: A polymer consisting of only one monomeric species.

HOPPER: Conical feed reservoir into which molding powder is loaded and from which it falls into a molding machine or extruder, sometimes through a metering device.

HOPPER DRYER: A combination feeding and drying device for extrusion and injection molding of thermoplastics. Hot air flows upward through the hopper containing the feed pellets.

HOT GAS WELDING: A technique of joining thermoplastic materials (usually sheet) where the materials are softened by a jet of hot air from a welding torch, and joined together at the softened points. Generally a thin rod of the same material is used to fill and consolidate the gap.

HOT-STAMPING: Engraving operation for marking plastics in which roll leaf is stamped with heated metal dies onto tne face of the plastics. Ink compounds can also be used. By means of felt rolls, ink is applied to type and by means of heat and pressure, type is impressed into the material, leaving the marking compound in the indentation.

HYDRAULIC: A system in which energy is transferred from one place to another by means of compression and flow of a fluid such as water, oil.

HYDROGENATION: Chemical process where hydrogen is introduced into a compound.

HYDROLYSIS: Chemical reaction of a substance with water.

HYGROSCOPIC: Tends to absorb moisture.

I

IMMISCIBLE: Descriptive of two or more fluids which are not mutually soluble.

IMPACT BAR (SPECIMEN): A test specimen of specified dimensions which is utilized to determine the relative resistance of a plastic to fracture as a result of shock.

IMPACT RESISTANCE: Relative susceptibility of plastics to fracture by shock, as indicated by the energy expended by a standard pendulum type impact machine in breaking a standard specimen in one blow.

IMPACT STRENGTH: (1) The ability of a material to withstand shock loading. (2) The work done in fracturing, under shock loading, a specified test specimen in a specified manner.

IMPREGNATION: The process of thoroughly soaking a material such as wood, paper or fabric, with a synthetic resin so the resin gets within the material.

IMPULSE SEALING: A heat sealing technique in which a pulse of intense thermal energy is applied to the sealing area for a very short time, followed immediately by cooling. It is usually accomplished by using a radio frequency heated metal bar which is cored for water cooling or is of such a mass that it will cool rapidly at ambient temperatures.

INHIBITOR: A substance that slows down a chemical reaction. Inhibitors are sometimes used in certain types of monomers and resins to prolong storage life.

INITIATOR: A substance that speeds up the polymerization of a monomer and becomes a component part of the chain.

INJECTION BLOW MOLDING: A blow molding process in which the parison to be blown is formed by injection molding.

INJECTION MOLD: A mold into which a material is introduced from an exterior heating cylinder.

INJECTION MOLDING: A molding procedure whereby a heat-softened plastic material is forced from a cylinder into a relatively cool cavity which gives the article the desired shape.

INSERT: An integral part of a plastics molding consisting of metal or other material which may be molded into position or may be pressed into the molding after the molding is completed.

INSERT, PROTRUDING: One having a part protruding from molded material.

INSERT, RIVET: One having a protruding part which is riveted in assembly.

INSULATION RESISTANCE: The electrical resistance of an insulating material to a direct voltage. It is determined by measuring the leakage of current which flows through the insulation.

INTERNAL MIXERS: Mixing machines consisting of cylindrical containers in which the materials are deformed by rotating blades or rotors. The containers and rotors are cored so that they can be heated or cooled to control the temperature of a batch. These mixers are extensively used in the compounding of plastics and rubber materials and have the inherent advantage of keeping dust and fume hazards to a minimum.

INVENTORY: In injection molding or extrusion, the amount of plastic contained in the heating cylinder or barrel.

IONOMER RESINS: Polmers which have ethylene as their major component, but containing both covalent and ionic bonds. The polymers exhibit very strong interchain ionic forces. The anions hang from the hydrocarbon chain and the cations are metallic-magnesium, zinc. These resins have many of the same features as polyethylene plus high transparency, tenacity, resilience and increased resistance to oils, greases and solvents. Fabrication is carried out as with polyethylene.

ISOCYANATE RESINS: Resins synthesized from isocyanates (- N = C = O) and alcohols (- OH). The reactants are joined through the formation

of the urethane linkage $\left(\begin{array}{c} O \\ \| \\ - N - C - O - \\ | \\ H \end{array} \right)$ and

hence this field of technology is generally known as urethane chemistry.

ISOTACTIC: A stereo regular structure of a vinyl polymer in which the side chains are aligned on the same side of the plane of the polymer backbone.

J

JET MOLDING: Processing technique characterized by the fact that most of the heat is applied to the material as it passes through the nozzle or jet, rather than in a heating cylinder as is done in conventional processes.

JIG: Tool for holding component parts of an assembly during the manufacturing process, or for holding other tools. Also called a Fixture.

JOINT: The location at which two adherends are held together with a layer of adhesive.

JOINT, BUTT: A type of edge joint in which the edge faces of the two adherends are at right angles to the other faces of the adherends. See JOINT; JOINT, EDGE; JOINT, SCARF.

JOINT, EDGE: A joint made by bonding the edge faces of two adherends with adhesives. See JOINT; JOINT, SCARF; JOINT, BUTT.

JOINT, LAP: A joint made by placing one adherend partly over another and bonding together the overlapped portions. See JOINT, SCARF.

JOINT, SCARF: A joint made by cutting away similar angular segments of two adherends and bonding the adherends with the cut areas fitted together. See JOINT, LAP.

JOINT, STARVED: A joint which has an insufficient amount of adhesive to produce a satisfactory bond. This condition may result from too thin a spread to fill the gap between the adherends, excessive penetration of the adhesive into the adherend, too short an assembly time, or the use of excessive pressure.

K

KISS-ROLL COATING: This roll arrangement carries a metered film of coating to the web; at the line of web contact, it is split with part remaining on the roll, the remainder of the coating adhering to the web.

KNIFE COATING: A method of coating a substrate (usually paper or fabric) in which the substrate, in the form of a continuous moving web, is coated with a material whose thickness is controlled by an adjustable knife or bar set at a suitable angle to the substrate. In the plastics industry PVC formulations are widely used in this work and curing is effected by passing the coated substrate into a special oven, usually heated by infrared lamps or convected air. There are a number of variations of this basic technique and they vary according to the type of product required.

KNOCKOUT PIN: A device for knocking a cured piece from a mold. Also called Ejector Pin.

L

LACQUER: Solutions of natural or synthetic resins, in readily evaporating solvents, which are used as protective coatings.

LAMINATED PLASTICS (SYNTHETIC RESIN-BONDED LAMINATE, LAMINATE): A plastics material consisting of superimposed layers of a synthetic resin-impregnated or resin-coated filler which have been bonded together, usually by means of heat and pressure, to form a single piece.

LAMINATED WOOD: A high-pressure bonded wood product composed of layers of wood with resin as the laminating agent. Plywood is a form of laminated wood in which successive layers of veneer are ordinarily cross laminated, the core of which may be veneer or sawed lumber in one piece or in a number of pieces.

LAND: (1) The horizontal bearing surface of a semipositive or flash mold by which excess material escapes. (2) The bearing surface along the top of the flights of a screw in a screw extruder. (3) The surface of an extrusion die parallel to the direction of melt flow.

LATCH: Device used to hold together two members of a mold.

LAY-UP: As used in reinforced plastics, the reinforcing material placed in position in the mold; also the resin-impregnated reinforcement.

LAY UP: The process of placing the reinforcing material in position in the mold.

LIGHT-RESISTANCE: The ability of a plastics material to resist fading after exposure to sunlight or ultraviolet light. Nearly all plastics tend to darken under these conditions.

L/D RATIO: A term used to define an extrusion screw which denotes the ratio of the screw length to the screw diameter.

LINEAR MOLECULE: A long chain molecule of two dimensional structure which may or may not contain side chains or branches. In effect, structural units connected to one another in a linear sequence.

LINTERS: Short fibers that adhere to the cotton seed after ginning. Used in rayon manufacture, as fillers for plastics, and as a base for the manufacture of cellulosic plastics.

LOADING TRAY (CHARGING TRAY): A device in the form of a specially designed tray which is used to load the charge simultaneously into each cavity of a multicavity mold by the withdrawal of a sliding bottom from the tray.

LOOSE DETAIL MOLD: A mold having parts that come out with the piece.

LOW-PRESSURE LAMINATES: In general, laminates molded and cured in the range of pressures from 400 psi down to and including pressures obtained by the mere contact of the plies.

LUCITE: A trade name of du Pont, also used generically for polymers prepared from methyl methacrylate.

LUG: An indentation or raised portion of the surface of a container, provided to control automatic (multicolor) decorating operations.

LUMINESCENT PIGMENTS: Special pigments available to produce striking effects in the dark. Basically there are two types; one is activated by ultraviolet radiation, producing very strong luminescence and, consequently, very eye-catching effects; the other type, known as phosphorescent pigments, does not require any separate source of radiation.

M

MACERATE: (v) To chop or shred fabric for use as a filler for a molding resin. (n) The molding compound obtained when so filled.

MACHINE SHOT CAPACITY: Refers to the maximum weight of a specific thermoplastic resin which can be displaced or injected by the injection ram in a single stroke.

MANIFOLD: A term used mainly with reference to blow molding and sometimes with injection molding equipment. It refers to the distribution or piping system which takes the single channel flow output of the extruder or injection cylinder and divides it to feed several blow molding heads or injection nozzles.

MAT: A randomly distributed felt of glass fibers used in reinforced plastics lay-up molding.

MATERIAL WELL: Space provided in a compression or transfer mold to care for bulk factor.

MATCHED METAL MOLDING: Method of molding reinforced plastics between two close-fitting metal molds mounted in a hydraulic press.

MECHANICALLY FOAMED PLASTIC: A cellular plastic whose structure is produced by physically incorporated gases.

MELAMINE FORMALDEHYDE RESIN: A synthetic resin derived from the reaction of melamine with formaldehyde.

MELT INDEX: The amount, in grams, of a thermoplastic resin which can be forced through a 0.0825 in. orifice when subjected to 2160 gms force in 10 minutes at 190°C.

MELT STRENGTH: The strength of the plastic while in the molten state.

MER: The repeating structural unit of any high polymer.

METALLIZING: Applying a thin coating of metal to a nonmetallic surface. May be done by chemical deposition or by exposing the surface to vaporized metal in a vacuum chamber.

METALLIC PIGMENTS: A class of pigments consisting of thin opaque aluminum flakes (made by ball milling either a disintegrated aluminum foil or a wrought metal powder and then polishing to obtain a flat, brilliant surface on each particle) or copper alloy flakes (known as bronze pigments). Incorporated into plastics, they produce unusual silvery and other metal-like effects.

METHYL METHACRYLATE:

$$CH_2 = C - COOCH_3$$ with CH_3 group

a colorless, volatile liquid derived from acetone cyanohydrin, methanol and dilute sulphuric acid, and used in the production of acrylic resins.

MODULES OF ELASTICITY: Stress/strain ratio in a material that is elastically deformed.

MOISTURE VAPOR TRANSMISSION: The rate at which water vapor permeates through a plastic film or wall at a specified temperature and relative humidity.

MOLD: (v) To shape plastic parts or finished articles by heat and pressure. (n) (1) The cavity or matrix into which the plastic composition is placed and from which it takes its form. (2) The assembly of all the parts that function collectively in the molding process.

MOLD BASE: The assembly of all parts making up an injection mold, other than the cavity, cores, and pins.

MOLDING CYCLE: (1) The period of time occupied by the complete sequence of operations on a molding press requisite for the production of one set of moldings. (2) The operations necessary to produce a set of moldings without reference to the time taken.

MOLDING POWDER: Plastic material in varying stages of granulation, and comprising resin, filler, pigments, plasticizers, and other ingredients, ready for use in the molding operation.

MOLDING PRESSURE: The pressure applied to the ram of an injection machine or press to force the softened plastic completely to fill the mold cavities.

MOLDING SHRINKAGE (MOLD SHRINKAGE, SHRINKAGE, CONTRACTION): The difference in dimensions, expressed in inches per inch, between a molding and the mold cavity in which it was molded, both the mold and the molding being at normal room temperature when measured.

MOLD RELEASE: See PARTING AGENT.

MOLECULAR WEIGHT DISTRIBUTION: A measure of the frequency of occurrence of the different molecular weight chains contained in a homologous polymeric system. The ratio of the weight average molecular weight to the number average molecular weight is sometimes used as an indication of the breadth of the distribution.

MONOMER: A relatively simple compound which can react to form a polymer. See also POLYMER.

MOUNTING PLATE: Part of blow molding unit to which the mold is attached.

MOVABLE PLATEN: The large back platen of an injection molding machine to which the back half of the mold is secured during operation. This platen is moved either by a hydraulic ram or a toggle mechanism.

MULTICAVITY MOLD: A mold with two or more mold impressions, for example, a mold which produces more than one molding per molding cycle.

MYLAR: A film produced from the polyester of ethylene glycol and terephthalic acid (dimethyl terephthalate actually used). The fiber made by this procedure is called Dacron. Mylar and Dacron are du Pont trade names.

N

NECK: The part of a container where the shoulder cross section area decreases to form the finish.

NEEDLE BLOW: A specific blow molding technique where the blowing air is injected into the hollow article through a sharpened hollow needle which pierces the parison.

NONPOLAR: Having no concentrations of electrical charge on a molecular scale (no dipole moment), thus, incapable of significant dielectric loss. Examples among resins are polystyrene and polyethylene.

NOZZLE: The hollow cored metal nose screwed into the extrusion end of (a) the heating cylinder of an injection machine or (b) a transfer chamber where this is a separate structure. A nozzle is designed to form, under pressure, a seal between the heating cylinder or the transfer chamber and the mold. The front end of a nozzle may be either flat or spherical in shape.

NYLON: The generic name for all synthetic fiber-forming polyamides; they can be formed into monofilaments and yarns characterized by great toughness, strength and elasticity, high melting point, and good resistance to water and chemicals. The material is widely used for bristles in industrial and domestic brushes, and for many textile applications; it is also used in injection molding gears, bearings, combs, etc.

O

OLEFINS: A group of unsaturated hydrocarbons of the general formula $C_n H_{2n}$, and named after the corresponding paraffins by the addition of "ene" or "ylene" to the stem. Examples are ethylene and pentene - 1.

ONE-SHOT MOLDING: In the urethane foam field, indicates a system whereby the isocyanate, polyol, catalyst, and other additives are mixed together directly and a foam is produced immediately (as distinguished from prepolymer).

OPEN-CELL FOAM: A cellular plastic in which there is a predominance of interconnected cells.

ORANGE-PEEL: Injection moldings that have

unintentionally rough surfaces.

ORGANOSOL: A resin (such as PVC) dispersion, the liquid phase of which contains one or more organic solvents. See also PLASTISOL.

ORIENTATION: The alignment of the crystalline structure in polymeric materials so as to produce a highly uniform structure. Can be accomplished by cold drawing or stretching during fabrication.

ORIFICE: The opening in the extruder die formed by the orifice bushings (ring) and mandrel.

OVERLAY SHEET (SURFACING MAT): A nonwoven fibrous mat (glass, synthetic fiber, etc.) used as the top layer in a cloth or mat lay-up to provide a smoother finish or minimize the appearance of the fibrous pattern.

P

PANELING: Distortion of a container occurring during aging or storage, caused by the development of a reduced pressure inside the container.

PARISON: The hollow plastic tube from which a container is blow molded.

PARISON SWELL: In blow molding, the ratio of the cross-sectional area of the parison to the cross-sectional area of the die opening.

PARTING AGENT: A lubricant such as wax or silicone oil used to coat a mold cavity to prevent the molded piece from sticking to it, and thus to facilitate its removal from the mold. Also called Release Agent.

PARTING LINE: Mark on a molding or casting where halves of mold meet in closing.

PARYLENE: Poly-para-xylene, used in ultra thin films for capacitor dielectrics and as a pore-free coating. Films are formed by heating a monomer and condensing it on a cool surface.

PERMANENT SET: The increase in length, expressed as a percentage of the original length, by which an elastic material fails to return to original length after being stressed for a standard period of time.

PERMEABILITY: (1) The passage or diffusion of a vapor, liquid, or solid through a barrier without physically or chemically affecting it. (2) The rate of such passage.

PHENOLIC RESIN: A synthetic resin produced by the condensation of phenol with formaldehyde in base. Phenolic resins form the basis of a family of thermosetting molding materials, laminated sheet, and stoving varnishes. They are also used as impregnating agents and as components of paints, varnishes, lacquers, and adhesives.

PHENOXY RESINS: A high molecular weight thermoplastic polyether resin based on bisphenol-A and epichlorohydrin having bisphenol-A terminal groups. Recently developed in the United States, the material is available in grades suitable for injection molding, extrusion, coatings and adhesives.

PHTHALATE ESTERS: A group of plasticizers produced by the direct action of alcohol on phthalic anhydride. They are the most widely used of all plasticizers, and are generally characterized by moderate cost, good stability, and good all-round properties.

PILL: See PREFORM.

PINCH-OFF: A raised edge around the cavity in the mold which seals off the part and separates the excess material as the mold closes around the parison in the blow molding operation.

PINCH-OFF TAIL: The bottom of the parison that is pinched off when the mold closes.

PINHOLE: A very small hole in an extruded resin coating of film.

PITCH: The distance from any point on the flight of a screw line to the corresponding point on an adjacent flight, measured parallel to the axis of the screw line or threading.

PLASTIC: (adj) Pliable and capable of being shaped by pressure. "Plastic" is incorrectly used as the generic word for the industry and its products.

PLASTICS TOOLING: Tools such as dies, jigs, fixtures, etc., for the metal forming trades constructed of plastics, generally laminates or casting materials.

PLASTICITY: The quality of being able to be shaped by plastic flow.

PLASTICIZE: To soften a material and make it plastic or moldable, either by adding a plasticizer or using heat.

PLASTICIZER: Chemical agents added to plastic compositions to improve flow and processability and to reduce brittleness. This is achieved by lowering the glass transition temperature.

PLASTISOLS: Mixture of resins and plasticizers which can be molded, cast, or converted to continuous films by the application of heat. If the mixtures contain volatile thinners also, they are known as Organisols.

PLATENS: The mounting plates of a press to which the entire mold assembly is bolted.

PLUG-AND-RING: Method of sheet forming in which a plug, functioning as a male mold, is forced into a heated plastic sheet held in place by a clamping ring.

PLUG FORMING: A thermoforming process in which a plug or male mold is used to partially preform the part before forming is completed using vacuum or pressure.

POLISHING ROLL(S): A roll or series of rolls, which has a highly polished chrome plated surface, that is utilized to produce a smooth surface on sheet as it is extruded.

POLYAMIDE: A polymer in which the structural units are lined by amide groupings

$$\left(\begin{array}{c} O \\ \| \\ - C - N - \\ | \\ H \end{array} \right).$$ Many polyamides are fiber-forming.

POLYBLENDS: A mechanical (nonchemical) mixture of two or more polymers, for example polystyrene and rubber.

POLYCARBONATE RESINS: Polymers derived from the direct reaction between aromatic and aliphatic dihydroxy compounds with phosgene or by the ester exchange reaction with appropriate phosgene derived precursors. Structural units are linked by carbonate groups

$$\left(\begin{array}{c} - O - C - O - \\ \| \\ O \end{array} \right).$$

POLYCONDENSATION: See CONDENSATION.

POLYESTER: A resin formed by the reaction between a dibasic acid and dihydroxy alcohol, both organic, or by the polymerization of a hydroxy carboxylic acid. Modification with multifunctional acids and/or alcohols and some unsaturated reactants permit cross-linking to thermosetting resins.

POLYETHYLENE: A thermoplastic material composed solely of ethylene. It is normally a translucent, tough, waxy solid which is unaffected by water and by a large range of chemicals.

POLYIMIDE RESINS: Aromatic polyimides made by reacting pyromellitic dianhydride with aromatic diamines. Characterized by high resistance to thermal stress. Applications include components for internal combustion engines.

POLYMER: A high molecular-weight organic compound, natural or synthetic, whose structure can usually be represented by a repeated small unit, the MER; for example polyethylene, rubber, cellulose. Synthetic polymers are formed by addition or condensation polymerization of monomers. Some polymers are elastomers, some plastics and some are fibers.

POLYMERIZATION: A chemical reaction in which the molecular weight of the molecule formed is a multiple of that of the original substances. When two or more monomers are involved, the process is called copolymeri-zation or heteropolymerization. See also DEGREE OF POLYMERIZATION, and POLYMER.

POLYMETHYL METHACRYLATE: A thermoplastic polymer synthesized from methyl methacrylate. It is a transparent solid with exceptional optical properties and good resistance to water. It is obtainable in the form of sheets, granules, solutions, and emulsions. It is extensively used for aircraft domes, lighting fixtures, decorative articles, etc.; it is also used in optical instruments and surgical appliances.

POLYPHENYLENE OXIDE: A polyether of 2, 6-dimethyl-phenol synthesized via an oxidative coupling process by means of air or pure oxygen in the presence of a copper-amine complex catalyst. These resins have a useful temperature range from less than -275 deg. F. to 375 deg. F. with intermittent use up to 400 deg. F. possible.

POLYPROPYLENE: A tough, lightweight, rigid plastic made by the polymerization of high-purity propylene gas in the presence of an organometallic catalyst at relatively low pressures and temperatures.

POLYSTYRENE: A water-white thermoplastic produced by the polymerization of styrene (vinyl benzene). The electrical insulating properties of polystyrene are outstandingly good and the material is relatively unaffected by moisture.

POLYSULFONE: A polymer containing the sulfone linkage $\left(\begin{array}{c} O \\ \| \\ - S - \\ \| \\ O \end{array} \right).$ These thermoplastic materials exhibit exceptional high temperature and low creep properties, have arc resistance, are self-extinguishing and may be molded and extruded.

POLYTETRAFLUOROETHYLENE (PTFE) RESINS: Members of the flurocarbon family of plastics made by the polymerization of tetrafluoroethylene. PTFE is characterized by its extreme inertness to chemicals, very high thermal stability and low frictional properties. Among the applications for these materials are bearings, fuel hoses, gaskets, tapes, and coatings for metal and fabric.

POLYURETHANE RESINS: A family of resins produced by reacting diisocyanates in excess with glycols to form polymers having free isocyanate groups. These groups, under the influence of heat or certain catalysts, will react with each other, or with water, glycols, etc., to form a thermoset.

POLYVINYL ACETATE: A thermoplastic material composed of polymers of vinyl acetate in the form of a colorless solid. It is obtainable in the form of granules, solutions, latices, and pastes, and is used extensively in adhesives, for paper and fabric coatings, and in bases for inks and lacquers.

POLYVINYL CHLORIDE (PVC): A thermoplastic polymer synthesized from vinyl chloride; a colorless solid with outstanding resistance to water, alcohols, and concentrated acids and alkalies. It is obtainable in the form of granules, solutions, and pastes. Compounded with plasticizers it yields a flexible material (plastisol) superior to rubber in aging properties. Widely used for cable and wire coverings and in making protective garments.

POLYVINYL CHLORIDE ACETATE: A thermoplastic copolymer of vinyl chloride and vinyl acetate; a colorless solid with good resistance to water, concentrated acids, and alkalies. It is obtainable in the form of granules, solutions, and emulsions. Widely used in protective garments and for cable and wire coverings. Compounded with plasticizers it yields a flexible material superior to rubber in aging properties.

POLYVINYLIDENE CHLORIDE: A thermoplastic polymer of vinylidene chloride. It is a white powder with softening temperature at 185-200 deg. C. The material is also supplied as a copolymer with acrylonitrile or vinyl chloride, giving products which range from the soft flexible type to the rigid type. Also known as Saran.

POSITIVE MOLD: A mold designed to trap all the molding material when it closes.

POSTFORMING: The forming, bending, or shaping of thermoset laminates that have been heated to make them flexible before the final thermosetting reaction has occurred. On cooling, the formed laminate retains the contours and shape of the mold over which it has been formed.

POT LIFE: See WORKING LIFE.

POTTING: Similar to Encapsulating, except that steps are taken to insure complete penetration of all the voids in the object before the resin polymerizes.

POWDER MOLDING: General term used to denote several techniques for producing objects of varying sizes and shapes by melting polyethylene powder, usually against the inside of a mold. The techniques vary as to whether the molds are stationary (such as in variations on slush molding techniques) or rotating (such as in variations on rotational molding).

PREFORM: (n) A compressed tablet or biscuit of plastic composition used for efficiency in handling and accuracy in weighing materials. (v) To make plastic molding powder into pellets or tablets.

PREHEATING: The heating of a compound prior to molding or casting in order to facilitate the operation or to reduce the molding cycle.

PREIMPREGNATION: The mixing of resin and reinforcing material before molding.

PREMIX: Materials in which the resin, reinforcement, extenders, fillers, etc., have been premixed before molding.

PREPOLYMER: A chemical intermediate whose molecular weight is between that of the monomer or monomers and the final polymer or resin.

PREPREG: A term generally used in reinforced plastics to mean the reinforcing material containing or combined with the full complement of resin before molding.

PRESS POLISH: A finish for sheet stock produced by contact, under heat and pressure, with a very smooth metal which gives the plastic a high sheen.

PRESSURE FORMING: A thermoforming process wherein pressure is used to push the sheet to be formed against the mold surface as opposed to using a vacuum to suck the sheet flat against the mold.

PRESSURE SENSITIVE ADHESIVE: An adhesive which develops maximum bonding power by applying only a light pressure.

PRIMER: A coating applied to a surface, prior to the application of a plastic coating, and adhesive or lacquer, enamel or the like, to improve the performance of the bond.

PROFILE DIE: An extrusion die for the production of continuous shapes, excepting tubes and sheets.

PROGRAMMING: The extrusion of a parison which differs in thickness in the length direction in order to equalize wall thickness of the blown container. It can be done with a pneumatic or hydraulic device which activates the mandrel shaft and adjusts the mandrel position during parison extrusion (parison programmer, controller, or variator). It can also be done by varying extrusion speed on accumulator-type blow molding machines.

PROMOTER: A chemical, itself a feeble catalyst, that greatly increases the activity of a given catalyst.

PROTOTYPE MOLD: A simplified mold construction often made from a light metal casting alloy or from an epoxy resin in order to obtain information for the final mold and/or part design.

PURGING: Cleaning one color or type of material from the cylinder of an injection molding machine or extruder by forcing it out with the new color or material to be used in subsequent production. Purging materials are also available.

PVC: Polyvinyl chloride.

Q

QUENCH (THERMOPLASTICS): A process of shock cooling thermoplastic materials from the molten state.

QUENCH BATH: The cooling medium used to quench molten thermoplastic materials to the solid state.

QUENCH-TANK EXTRUSION: Extrusion of a film cooled in a quench-water bath.

R

RADIO FREQUENCY (rf) PREHEATING: A method of preheating used for molding materials to facilitate the molding operation or reduce the molding cycle. The frequencies most commonly used are between 10 and 100 mc/sec.

RADIO FREQUENCY WELDING: A method of welding thermoplastics using a radio frequency field to apply the necessary heat. Also known as High Frequency Welding.

RANDOM COPOLYMER: Copolymer in which the molecules of each monomer are randomly arranged in the polymer backbone.

RAYON: The generic term for fibers, staple, and continuous filament yarns composed of regenerated cellulose. Rayon fibers are similar in chemical structure to natural cellulose fibers such as cotton except that the synthetic fiber contains shorter polymer units. Most rayon is made by the viscose process.

RECIPROCATING SCREW: An extruder system in which the screw when rotating is pushed backwards by the molten polymer which collects in front of the screw. When sufficient material has been collected, the screw moves forward and forces the material through the head and die at a high speed.

RECYCLE: Ground material from flash and trimmings which after mixing with a certain amount of virgin material is fed back into the blow molding machine.

REGENERATED CELLULOSE (CELLOPHANE): A transparent cellulose plastics material made by mixing cellulose xanthate with a dilute sodium hydroxide solution to form a viscose. Regeneration is carried out by extruding the viscose, in sheet form, into an acid bath to create regenerated cellulose. The material is very widely used as a packaging and overwrapped material of exceptional clarity. The film also has good electrical properties and

is resistant to oils and greases. Included among recent applications is the use of the material as a release agent in reinforced plastics molding.

REINFORCED MOLDING COMPOUND: Compound containing resin and a reinforcing filler, supplied in the form of ready-to-use materials; as distinguished from premix.

RELEASE AGENT: See PARTING AGENT.

RESIN: Any of a class of solid or semisolid organic products of natural or synthetic origin, generally of high molecular weight, with no definite melting point. Most resins are polymers.

RESISTIVITY: The ability of a material to resist passage of electrical current either through its bulk or on a surface. The unit of volume resistivity is the ohm-cm, or surface resistivity, the ohm.

RESTRICTED GATE: A very small orifice between runner and cavity in an injection or transfer mold. When the piece is ejected, this gate breaks cleanly, simplifying separation of runner from piece.

RHEOLOGY: The study of flow of polymeric materials on a macroscopic and microscopic level.

RIB: A reinforcing member of a fabricated or molded part.

RIGID PVC: Polyvinyl chloride or a polyvinyl chloride acetate copolymer characterized by a relatively high degree of hardness; it may be formulated with or without a small percentage of plasticizer.

ROCKWELL HARDNESS: A common method of testing a plastics material for resistance to indentation in which a diamond or steel ball, under pressure, is used to pierce the test specimen. The load used is expressed in kilograms and a 10-kilogram weight is first applied and the degree of penetration noted. The so-called major load (60 to 150 kilograms) is next applied and a second reading obtained. The hardness is then calculated as the difference between the two loads and expressed with nine different prefix letters to denote the type of penetrator used and the weight applied as the major load.

ROLLER COATING: Used for applying paints to raised designs or letters.

ROTATIONAL CASTING (OR MOLDING): A method used to make hollow articles from plastisols and latices. Plastisol is charged into a hollow mold capable of being rotated in one or two planes. The hot mold fuses the plastisol into a gel after the rotation has caused it to cover all surfaces. The mold is then chilled and the product stripped out.

RUNNER: In an injection of transfer mold, the channel, usually circular, that connects the sprue with the gate to the mold cavity.

S

SAN: Styrene acrylonitrile thermoplastic co-polymer with good stiffness, scratch, chemical, and stress crack resistance.

SANDWICH CONSTRUCTIONS: Panels composed of a lightweight core material, honeycomb, foamed plastic, etc., to which two relatively thin, dense, high strength faces or skins are adhered.

SATURATED COMPOUNDS: Organic compounds which do not contain double or triple bonds and thus cannot add on elements or compounds.

SCRAP: Any product of a molding operation that is not part of the primary product. In compression molding, this includes flash, culls, runners, and is not reusable as a molding compound. Injection molding and extrusion scrap (runners, rejected parts, sprues, etc.) can usually be reground and remolded.

SELF-EXTINGUISHING: A somewhat loosely-used term describing the ability of a material to cease burning once the source of flame has been removed.

SEMIAUTOMATIC MOLDING MACHINE: A molding machine in which only part of the operation is controlled by the direct action of a human. The automatic part of the operation is controlled by the machine according to a predetermined program.

SET: To convert a liquid resin or adhesive into a solid state by curing or by evaporation of solvent or suspending medium or by gelling.

SETTING TEMPERATURE: The temperature to which a liquid resin, an adhesive, or products or assemblies involving either must be heated in order to set them.

SETTING TIME: The period of time during which a molded or extruded product, an assembly, etc., is subjected to heat and/or pressure to set the resin or adhesive.

SHEAR STRENGTH: (1) The ability of a material to withstand shear stress. (2) The stress at which a material fails in shear.

SHEET (THERMOPLASTICS): A flat section of thermoplastic resin with the length considerably greater than the width and 10 mils or greater in thickness.

SHORE HARDNESS: A method of determining the hardness of a plastic material. The device used consists of a small conical hammer fitted with a diamond point and acting in a glass tube. The hammer is made to strike the material under test and the degree of rebound is noted on a graduated scale. Generally, the harder the material the greater will be the rebound.

SHORT OR SHORT SHOT: In injection molding, failure to fill the mold completely.

SHOT: The yield from one complete molding cycle, including scrap.

SHOT CAPACITY: The maximum weight of material which an accumulator can push out with one forward stroke of the ram.

SHRINKAGE: Contraction of a molded material upon cooling, or of a casting upon polymerizing.

SILK SCREEN PRINTING (SCREEN PROCESS DECORATING): This printing method, in its basic form, involves laying a pattern of an insoluble material, in outline, on a finely woven fabric, so that when ink is drawn across it, it is able to pass through the screen only in the desired areas.

SINGLE CAVITY MOLD (INJECTION): An injection mold having only one cavity in the body of the mold, as opposed to multiple cavity molds or family molds which have numerous cavities.

SILICONE: One of the family of polymeric materials in which the recurring chemical group contains silicon and oxygen atoms as links in the main chain. At present these compounds are derived from silica (sand) and methyl chloride. The various forms obtainable are characterized by their resistance to heat and low coefficients of thermal expansion. Silicones are used in the following applications: (a) Greases for lubrication. (b) Rubber-like sheeting for gaskets, etc. (c) Heat-stable fluids and compounds for waterproofing, insulating, etc. (d) Thermosetting insulating varnishes and resins for both coating and laminating.

SINK MARK: A shallow depression or dimple on the surface of an injection molded part due to collapsing of the surface following local internal shrinkage after the gate seals. May also be an incipient short shot.

SINTERING: In forming articles from fusible powders such as nylon, the process of holding the pressed-powder article at a temperature just below its melting point for a period of time. Particles are fused (sintered) together, but the mass, as a whole, does not melt.

SLIP FORMING: Sheet forming technique in which some of the plastic sheet material is allowed to slip through the mechanically operated clamping rings during a stretch-forming operation.

SLUSH MOLDING: Method for casting thermoplastics, in which the resin in liquid form is poured into a hot hollow mold where a viscous skin forms. Excess slush is drained off, the mold is cooled, and the molding stripped out.

SNAP-BACK FORMING: Sheet forming technique in which an extended heated plastic sheet is allowed to contract over a form shaped to the desired contours.

SOFTENING RANGE: The range of temperature in which a plastic changes from a rigid to a soft state. Actual values will depend on the method of test. Sometimes erroneously referred to as softening point.

SOLUTION: Homogeneous mixture of two or more components, for example, gas dissolved in gas or liquid, or a solid in a liquid.

SOLVENT: Any substance, usually a liquid, which dissolves other substances.

SOLVENT MOLDING: Process for forming thermoplastic articles by dipping a mold in a solution or dispersion of the resin and drawing off the solvent to leave a layer of plastic film adhering to the mold.

SPANISHING: A method of depositing ink in the valleys of embossed plastics film.

SPECIFIC GRAVITY: The density (mass per unit volume) of any material divided by the density of water at a standard temperature, usually 4 deg. C. Since water's density is nearly 1.00 g/cc, density in g/cc and specific gravity are numerically nearly equal.

SPECIFIC HEAT: The amount of heat required to raise a specified mass by one unit of a specified temperature.

SPIN WELDING: A process of fusing two objects together by forcing them together while one of the pair is spinning, until frictional heat melts the interface. Spinning is then stopped and pressure held until they are frozen together. Also called INERTIA WELDING.

SPIRAL FLOW TEST: A method for determining the flow properties of a thermoplastic resin in which the resin flows along the path of a spiral cavity. The length of the material which flows into the cavity and its weight gives a relative indication of the flow properties of the resin.

SPLIT CAVITY: Cavity made in sections.

SPRAY COATING: Usually accomplished on continuous webs by a set of reciprocating spray nozzles traveling laterally across the web as it moves.

SPRAYED METAL MOLDS: Mold made by spraying molten metal onto a master until a shell of predetermined thickness is achieved. Shell is then removed and backed up with plaster, cement, casting resin, or other suitable material. Used primarily as a mold in sheet-forming.

SPRAY-UP: Covers a number of techniques in which a spray gun is used as the processing tool. In reinforced plastics, for example, fibrous glass and resin can be simultaneously deposited in a mold. In essence, roving is fed through a chopper and ejected into a resin stream which is directed at the mold by either of two spray systems. In foamed plastics, very fast-reacting urethane foams or epoxy foams are fed in liquid streams to the gun and sprayed on the surface. On contact, the liquid starts to foam.

SPREADER: A streamlined metal block placed in the path of flow of the plastics material in the heating cylinder of extruders and injection molding machines to spread it into thin layers, thus forcing it into contact with heating areas.

SPRUE: Feed opening provided in the injection or transfer mold; also the slug formed at this hole. Spur is a shop term for the sprue slug.

SPRUE BUSHING: A hardened steel insert in an injection mold which contains the tapered sprue hole and has a suitable seat for the nozzle of the injection cylinder. Sometimes called an Adapter.

STABILIZER: An ingredient used in the formulation of some polymers to assist in maintaining the physical and chemical properties of the compounded materials at their initial values throughout the processing and service life of the material, for example heat and ultraviolet stabilizers.

STATIONARY PLATEN: The large front plate of an injection molding machine to which the front plate of the mold is secured during operation. This platen does not move during normal operation.

STEAM MOLDING (EXPANDABLE POLYSTYRENE): Used to mold parts from pre-expanded beads of polystyrene using steam as a source of heat to expand the blowing agent in the material. The steam in most cases is contacted intimately with the beads directly or may be used indirectly to heat mold surfaces which are in contact with the beads.

STEREOSPECIFIC POLYMERS: Implies a specific or definite order of arrangement of molecules in space. This ordered regularity of the molecules in contrast to the random arrangement of molecules permits close packing of the molecules and leads to high crystallinity (as in isotactic polypropylene).

STRESS CRACK: External or internal crack in a plastic caused by tensile stresses. The development of such cracks is frequently accelerated by the environment to which the plastic is exposed. The stresses which cause cracking may be present internally or externally or may be combinations of these stresses. The appearance of a network of fine cracks is called crazing.

STRETCH FORMING: A plastic sheet forming technique in which the heated thermoplastic sheet is stretched over a mold and subsequently cooled.

STRIPPER-PLATE: A plate that strips a molded piece from core pins or force plugs. The stripper-plate is set into operation by the opening of the mold.

STYRENE PLASTICS: Plastics made by the polymerization of styrene or copolymerization of styrene with other unsaturated compounds.

SURFACE TREATING: Any method of treating a polymer so as to alter the surface rendering it receptive to inks, paints, lacquers, adhesives and processes such as chemical, flame, and electronic treating.

SUSPENSION: A mixture of fine particles of any solid within a liquid or gas. The particles are called the dispersed phase, the suspending medium is called the continuous phase.

SYNDIOTACTIC: A vinyl polymer in which the side chains alternate regularly above and below the plane of the backbone.

T

TACK: Stickiness of an adhesive, measurable as the force required to separate an adherend from it by viscous or plastic flow of the adhesive.

TENSILE BAR (SPECIMEN): A compression or injection molded specimen of specified dimensions which is used to determine the tensile properties of a material.

TENSILE STRENGTH: The pulling stress, in psi, required to break a given specimen. Area used in computing strength is usually the original, rather than the necked-down area.

TERPOLYMER: A polymeric system which contains three monomeric units, for example ABS (acrylonitrile, butadiene, styrene) terpolymer.

THERMAL CONDUCTIVITY: Ability of a material to conduct heat; quantity of heat that passes through a unit cube of a substance in unit of time when difference in temperature of two faces is one degree.

THERMAL STRESS CRACKING (TSC): Crazing and cracking of some thermoplastic resins which results from over-exposure to elevated temperatures.

THERMOFORMING: Any process of forming thermoplastic sheet which consists of heating the sheet and pulling it down onto a mold surface.

THERMOFORMS: The products which result from completing a thermoforming operation.

THERMOPLASTIC: (adj) Capable of being repeatedly softened by heat and hardened by cooling. (n) A material which has a linear macromolecular structure that will repeatedly soften when heated and harden when cooled. Typical of the thermoplastics family are the styrene polymers, and copolymers, acrylics, cellulosics, polyethylenes, vinyls, nylons, and the various fluorocarbon materials.

THERMOSET: A material that will undergo or has undergone a chemical reaction by the action of heat, catalysts, ultraviolet light, etc., leading to a relatively infusible and cross-linked stated. Typical of the plastics in the thermosetting family are the epoxies, melamines, urea formaldehyde resins, and phenolics.

THIXOTROPIC: Said of materials that are gel-like at rest but fluid when agitated. Liquids containing suspended solids are apt to be thixotropic. Thixotropy is desirable in paints.

TRANSFER MOLDING: A method of molding thermosetting materials, in which the plastic is first softened by heat and pressure in a transfer chamber, then forced at high pressure through suitable sprues, runners, and gates into closed mold for final curing.

TUMBLING: Finishing operation for small plastic articles by which gates, flash, and fins are removed and/or surfaces are polished by rotating them in a barrel together with wooden pegs, sawdust, and polishing compounds.

U

ULTRASONIC SEALING: A film sealing method in which sealing is accomplished through the application of vibratory mechanical pressure at ultrasonic frequencies (20 to 40 kc). Electrical energy is converted to ultrasonic vibrations through the use of either a magnetostrictive or piezoelectric transducer. The vibratory pressures at the film interface in the sealing area develop localized heat losses which melt the plastic surfaces effecting the seal.

UNDERCUT: (adj) Having a protuberance or indentation that impedes withdrawal from a two-piece, rigid mold. Flexible materials can be ejected intact even with slight undercuts. (n) Any such protuberance or indentation; depends also on design of mold.

UNSATURATED COMPOUNDS: Any compound having more than one bond between two adjacent atoms, usually carbon atom, and capable of adding other atoms at that point to reduce it to a single bond.

UREA FORMALDEHYDE RESIN (UREA RESIN): A synthetic thermoset resin derived from the reaction of urea (carbamide) with formaldehyde or its polymers.

URETHANE: See ISOCYANATE RESINS.

UV STABILIZER (ULTRAVIOLET): Any chemical compound which, admixed with a thermoplastic, selectively absorbs UV rays.

V

VACUUM FORMING: Method of sheet forming in which the plastic sheet is clamped in a stationary frame, heated, and drawn down by a vacuum into a mold. In a loose sense, it is sometimes used to refer to all sheet forming techniques, including Drape Forming, involving the use of vacuum and stationary molds.

VACUUM METALLIZING: Process in which surfaces are thinly coated with metal by exposing them to the vapor of metal that has been evaporated under vacuum.

VALLEY PRINTING: Ink is applied to the high points of an embossing roll and subsequently deposited in what becomes the valleys of the embossed plastic material.

VINYL RESIN: A synthetic resin formed by the polymerization of chemical compounds containing the group $CH_2 = CH -$. For example, polyvinyl chloride, acetate, alcohol, and butyral.

VISCOSITY: Internal friction or resistance to flow of a liquid. The constant ratio of shearing stress to rate of shear. In liquids for which this ratio is a function of stress, the term "apparent viscosity" is defined as this ratio.

VULCANIZATION: The chemical reaction which induces extensive changes in the physical properties of a rubber or plastic and which is brought about by reacting it with sulphur and/or other suitable agents. The changes in physical properties include decreased plastic flow, reduced surface tackiness, increased elasticity, much greater tensile strength, and considerably less solubility.

W

WARPAGE: Dimensional distortion in a plastic object after molding.

WEB: Thin sheet in process in machine. Molten web is that which issues from the die. Substrate web is the substrate being coated.

WELD LINES: A mark on a container caused by incomplete fusion of two streams of molten polymer.

WIRE TRAIN: The entire assembly which is utilized to produce a resin-coated wire which normally consists of an extruder, a crosshead and a die, cooling means, and feed and take-up spools for the wire.

WORKING LIFE: The period of time during which a liquid resin or adhesive, after mixing with catalyst, solvent, or other ingredients, remains usable.

Y

YIELD VALUE (YIELD STRENGTH): The lowest stress at which a material undergoes plastic deformation. Below this stress, the material is elastic; above it, viscous.

ACKNOWLEDGMENTS

The authors wish to express their appreciation to the many manufacturing companies and individuals who provided valuable information, literature, and photographs for the preparation of this text.

Special credit is given to Marge Baird and Mary Baird for their patience and encouragement during the preparation of the manuscript and illustrations.

Contributions from the following companies and organizations were sincerely appreciated.

AAA Plastics Equipment Co., Fort Worth, Texas
Acromark Co., Berkeley Heights, N.J.
Adamson United Co., Akron, Ohio
Allied Chemical Corp., New York, N.Y.
Alma Plastics Co., Alma, Mich.
Alpha Chemical and Plastics Corp., Newark, N.J.
Alpine American Corp., Natick, Mass.
American Cyanamid Co., Wayne, N.J.
American Hoechst Corp., Leominster, Mass.
Amoco Chemicals Corp., Chicago, Ill.
Arco Polymers, Inc., Philadelphia, Pa.
Battenfeld Corp. of America, Chicago, Ill.
Binks Manufacturing Co., Franklin Park, Ill.
Black Clawson Co., Fulton, N.Y.
Borg-Warner Chemicals, Parkersburg, W. Va.
Branson Sonic Power Co., Danburg, Conn.
Brown Machine Co., Beaverton, Mich.
The Budd Co., Madison Heights, Mich.
Cadillac Plastic and Chemical Co., Birmingham, Mich.
Celanese Plastics & Specialties Co., Newark, N.J.
Chemplast Inc., Wayne, N.J.
Chemplex Co., Rolling Meadows, Ill.
Cincinnati Milacron Co., Batavia, Ohio
Cosden Oil and Chemical Co., Big Spring, Texas
Dake Corp., Grand Haven, Mich.
Dow Chemical Co., Midland, Mich.
Dow Corning Corp., Midland, Mich.
E.I. du Pont de Nemours & Co., Wilmington, Del.
Eastman Chemical Products, Inc., Kingsport, Tenn.
Farrel Co., Rochester, N.Y.
Fellows Gear Shaper Co., Springfield, Vt.
Fiberfil Div., Dart Industries Inc., Evansville, Ind.
Filmaster Design Inc., Fairfield, N.J.
FMC Corp., Green Bay, Wisc.
Foremost Machine Builders, Inc., Fairfield, N.J.
Formica Corp., Wayne, N.J.
Franklin Manufacturing Corp., Norwood, Mass.
Gatto Machinery Development Corp., Hauppauge, N.Y.
General Electric Co., Pittsfield, Mass.
Gloucester Engineering Co., Inc., Gloucester, Mass.
B.F. Goodrich Co., Cleveland, Ohio
Gulf Oil Chemicals Co., Orange, Texas
Hayes-Albion Corp., Bay City, Mich.
Hayssen Manufacturing Co., Sheboygan, Wisc.
Honeywell Inc., Fort Washington, Pa.
Hooker Chemicals and Plastics Corp., Niagara Falls, N.Y.
Hoover Universal Inc., Ann Arbor, Mich.
Hull Corp., Hatboro, Pa.
ICI Americas Inc., Wilmington, Del.

Inta-Roto, Inc., Richmond, Va.
Kamweld Products Co., Norwood, Mass.
Killion Extruders, Inc., Verona, N.J.
Kissam Manufacturing, Inc., Mountainside, N.J.
Laramy Products Co., Inc., Lyndonville, Vt.
Lexington United Corp., St. Louis, Mo.
Mastersonics, Inc., Granger, Ind.
McNeil Akron Corp., Akron, Ohio
Michigan Chrome and Chemical Co., Detroit, Mich.
Michigan Oven Co., Romulus, Mich.
Mobay Chemical Corp., Pittsburg, Pa.
Molded Fiber Glass Companies, Ashtabula, Ohio
National Automatic Tool Co., Richmond, Ind.
New Britain Plastics Machine Co., New Britain, Conn.
Newbury Industries Inc., Newbury, Ohio
Northern Petrochemical Co., Des Plaines, Ill.
Owens-Corning Fiberglas Corp., Toledo, Ohio
Package Machinery Co., East Longmeadow, Mass.
Pennwalt Stokes Corp., Philadelphia, Pa.
Polymer Machinery Corp., Berlin, Conn.
Rexene Co., Paramus, N.J.
Rextrusion Systems, Inc., Saddle Brook, N.J.
Rislau Corp., Glen Rock, N.J.
W.S. Rockwell Co., Fairfield, Conn.
Rogers Corp., Rogers, Conn.
Rohm and Haas Co., Philadelphia, Pa.
A. Schulman Inc., Troy, Mich.
Scott Machine Development Corp., Walton, N.Y.
Service Textonics Inc., Adrian, Mich.
Shell Chemical Co., Houston, Texas
Snider Mold Co., Mequon, Wisc.
Sterling Extruder Corp., South Plainfield, N.J.
Stevens Elastomeric and Plastic Products Co., Easthampton, Mass.
Stewart Bolling and Co., Cleveland, Ohio
Testing Machines Inc., Amityville, N.Y.
Tinius Olsen Testing Machine Co., Willow Grove, Pa.
Ultra Sonic Seal Co., Broomall, Pa.
Union Carbide Corp., New York, N.Y.
Uniroyal Chemical, Naugatuk, Conn.
U.S. Industrial Chemicals Co., New York, N.Y.
U.S. Steel Corp., Pittsburg, Pa.
Vertrod Corp., Brooklyn, N.Y.
Wabash Metal Products Co., Wabash, Ind.
Wayne Machine and Die Co., Totowa, N.J.
Welex Inc., Blue Bell, Pa.
Wellman, Inc., Boston, Mass.
Westlake Plastics Co., Lenni, Pa.
Williams-White and Co., Moline, Ill.

INDEX

Index